Journeywoman

swinging a hammer in a man's world

An injury to one is
an injury to all!

Kate Braid

CAITLIN PRESS

More books by Kate Braid

Poetry

To This Cedar Fountain (1995, reprinted 2012)
Inward to the Bones: Georgia O'Keeffe's Journey with Emily Carr
(1998, reprinted 2010)
Turning Left to the Ladies (2009)
A Well-Mannered Storm: The Glenn Gould Poems (2008)
In Fine Form: The Canadian Book of Form Poetry,
Co-edited with Sandy Shreve (2005, reprinted 2006)
Covering Rough Ground (1991)

Non-fiction

Emily Carr: Rebel Artist (2000, 2007, 2011, reprinted in Japanese 2009)
The Fish Come in Dancing:
Stories from the West-Coast Fishery Editor (2002)
Red Bait! Struggles of a Mine Mill Local
Co-written with Al King (1998)

Every act of becoming conscious...is an unnatural act.
—Adrienne Rich, "The Phenomenology of Anger"

I'm frightened all the time. Scared to death.
But I've never let it stop me. Never!
—Georgia O'Keeffe

Does the flower know

Does it find itself a meadow
And open to whatever comes

And open to whatever comes
And open to whatever comes
—Marilyn Bowering, "Satin Flower"

This book is for all the blue collar women

especially in memory of Jacqueline Frewin, carpenter,
Carlyal Gittens, tile-setter and brick layer,
and my two aunts, Kathleen Coates and Kelly Pryde,
"Rosies" both

for my mother, Mary Elizabeth Braid
and in memory of my father, Harry Braid

with love.

Contents

1

dancing with lumber—11

2

madness and metaphor—23

3

pender island—37

4

head down, ass in the air—52

5

the best man for the job—69

6

contradictions—84

7

like sewing—93

8

slippery, sloping surface—102

9

shameless—114

10

union maid meets concrete—127

11

exotic dancing—144

12

how is a woman like a hammer?—167

13

doing it right—184

14

the beauty of ply—197

15

right/left/left/right—211

16

dease lake—225

17

finding gold—240

18

what's in the flesh—259

acknowledgements & thanks—267

photo credits—271

1

dancing with lumber

"Watch out!"

Someone jerks my arm—hard—and as I catch myself, I glimpse Janet's mouth shaped into a perfect soundless scream. Beside her, the forklift operator screeches to a halt, his back tire inches from me. His face mimes surprise and apology as Tom lets go of my arm and gestures for me to flip up my brand new ear muffs. I can hardly hear over the screech of saws.

"Watch out for the forklift!"

I nod, shaken, as Tom carries on, pointing up to where a set of rattling black chains, wide as a bed, divides above us. I feel thunder in my chest—the clash and groan of metal, scrape of lumber, howl of saws—and a smell so sharp and strong, it's like walking through Buckley's Cough Medicine. Soon, I'll learn it's the smell of thousands of board feet of freshly cut lumber.

"Grade A material," Tom is screaming into my exposed ear, "stays up top!" He waves at a guy pushing buttons on a machine above us and to our left. I order myself to pay attention; he's telling us how to do this job. "Rejects down here!" A small white feminine hand appears out of the plywood shack above us and to our right. "Grader's shack!" Tom screams. The hand pushes logs off the top chain down to the lower one, where they fall in a clatter in front of Janet and me. Our job will be to pile these rejects on carts so the forklift can load them into boxcars.

This should be easy.

Janet and I had met a few months earlier, at an International Women's Day celebration in Vancouver. When someone had asked about my plans for fall, I said I wanted to go back to university and for that, I needed a chunk of money, fast. When they asked, "How are you going to get it?" I said, "Up north." Actually, until the words came out of my mouth, I had no plan at all, but in 1975, whenever a guy wanted money in British Columbia, he went "up north"—to Kitimat, Smithers, Prince George—and came back with pocketfuls. It was boom days in northern BC; trees were being cut as fast as loggers could topple them, and dams, roads, mills, whole towns were popping up. If a man could earn big money up north, I suddenly thought, why couldn't I?

A British accent over my left shoulder chimed, "Me too." It was Janet, and I was in luck. She'd taught in BC's north for two years, and exotic words like Telkwa, Nass, Kitwanga, now rippled off her tongue as she told me the names of people she knew, places we could stay. Best of all, the words included places we might work. "Sawmill," she said. "Fish plant. Pulp mill." It never crossed my mind to ask if they hired women in sawmills. Instead, I phoned my mum and dad to tell them I was going north this summer to work. I was vague about the details.

We shopped at an army surplus store on East Broadway for what Janet called "necessities": plates, cutlery, a wallet-sized silver blanket ("Developed by NASA for Outer Space!"), sleeping bags, steel-toed boots and, for me, an ancient Trapper Nelson pack, the kind made of green canvas with a wood frame and leather straps. A few weeks later, Trapper Nelson at my side, Janet and I were standing at the on-ramp to Highway 1 with our thumbs out.

The world looks bigger from the back of a pickup truck. Heading east from Vancouver, past the town of Hope, then north, the country was no longer the steep up-and-down, brilliant greens of the west coast; it was long flat rolling hills, soft browns and the smell of something sweet.

"Sage," Janet said.

The pickup dropped us in downtown Williams Lake. I was busy ogling store windows full of cowboy boots and Stetson hats—as foreign as haute couture in Paris—when an old man walked by.

"Howdy."

They even spoke a foreign language. There was a sawmill in this town, and Janet used the restaurant phone to ask if they needed any help. Her face told me the answer. We picked up our packs, headed back to the highway, and for the next two weeks applied at every sawmill, paper mill and fish processing plant between Williams Lake and Prince Rupert. After twenty-eight years of living in cities and working in offices, it felt odd to talk to bosses called foremen who wore blue jeans instead of a suit. After each foreman took one look and said, "Sorry, girls," we moved on to the next. But when we got to Prince Rupert, to the coast, we'd turned around and started back toward Prince George in the diminishing hope that we'd somehow missed a place, that one of those foremen might change his mind. So far we'd only spent money, and it was almost June.

Finally, a truck driver told us he'd heard Canadian National Rail was hiring line crew at Hazelton, to lay track. We had no idea what "laying track" meant, but by now we were desperate. The trucker let us off at the Hazelton exit and pointed north. Janet and I walked up a gravel track and reached a trailer marked CNR just as a man pulling on his baseball cap stepped outside. As he brushed past he said amiably, "Looking for work, girls?"

"Yes."

That stopped him. As we shook hands it was suddenly all I could do not to turn and run. It felt weird to be standing outside while we talked about work, and what if this wild plan was about to actually become real? The foreman said he'd never hired a woman before, and even if he wanted to, he had no accommodation for us. When Janet prompted him, "You can put us in with the cooks," he pulled the brim of his cap back and forth across his forehead a few times, then gave us an even look.

"You know what this job is, don't you?"

We had no idea and he knew it, so he told us. We'd be going miles into the bush to tear up and lay down track and would be gone two, three weeks—us and a train full of men. He finished, "You'll wreck your hands."

But this was the only job we'd had even a whiff of and I was into it now. I boasted we were strong, we'd get stronger, we'd wear gloves.... He said he'd think about it and told us to meet him at 4:30 in the pub. We waited until 5:00, hoping the beer might soften him up.

"No."

"Why not?"

"The cooks you're so eager to stay with are pushing fifty and can upend a drunken road hand better than I can. I won't take two young women out into the bush with a crew of rangytangs because once the boys get into the liquor I'll end up spending my nights camped in front of your door with a rifle across my knees."

That did it for me, but Janet fired a last volley, threatening to go to the Canadian Human Rights Commission, like the women who'd just launched a law suit against CN in Quebec for exactly this reason.

"Sue me," he said kindly. "I won't hire you."

And that was it. We both knew there was virtually no chance of finding work before we reached Prince George that evening, and in that case, we might as well go home. A few rides later, less than seventy miles before Prince George, an Indo–Canadian couple with a child in the back seat picked us up. It was the woman who said, "There's work at the mill in Fort St. James." She knew because she used to work there herself. Her name was Parminder, and that night we squeezed into the spare bedroom in her family's tiny trailer.

The next morning, Janet marched ahead of me across a huge parking lot filled with cut lumber and whole logs and that Buckley's smell toward a small trailer with the sign Apollo Forest Products above its door, where a man behind a desk told us yes, he needed people for work in his planer mill, starting immediately.

"Pays $5.25 an hour."

Janet and I were both nodding as he handed us each a pair of leather gloves, a red hard hat and a set of yellow ear muffs. The last job I'd had— one of those federal Opportunities for Youth jobs that kept unemployed young people off welfare—paid $3.

The man's name was Archie and he was the owner. The hard hat felt rigid and unbalanced, as if someone had plopped a cardboard box on my forehead. Archie showed us how to attach the muffs, calling them "Mickey Mouse ears," then led us across the yard to a low industrial building with corrugated metal sides. When he opened a side door, the screech of metal made us clutch at our ears as he pointed furiously at our heads.

"Ear muffs!"

We pulled them down awkwardly but lifted them again so he could scream, "Tom! Your foreman!" when a small man with a blue hard hat, muffs up, appeared before us.

Tom led Janet and me inside, along the long arm of the double chain that ground and screeched as sticks of wood bobbed over it like logs on water. I caught a glimpse of men with their backs to us, monster saws, then a sharp right to where a railway car sat open to our left, half-packed with lumber. It all felt surreal, bigger than it should be, as if—like Alice into Wonderland—stepping through the door of the mill had changed everything.

—

Now, after saving me from the forklift, Tom picks a piece of wood off the chain to show us the blue scribble on its end.

"Grader marking! Too short! A reject!" He fires it onto one of the carts lined up beside the chain and instantly grabs another. "Bark!" A log with tree bark still showing on one side and a different scribble that he fires onto a second cart.

The carts are metal beds on wheels with a wall at the back, and the *thonk*! of wood against steel adds to the din. He wants us to pile these rejects, each on the right cart depending on its scribble, dividing the layers with spacers—"Suckers!" he yells in my ear, and for a second I think he's referring to Janet and me. Then, nodding for Janet to stand at the front of this section and me at the back, he gives us a thumbs-up and is gone.

I spent years learning to type sixty words a minute, but any time I ever applied for a secretarial job, they took me into an airless cell, told me to relax and gave me a typing test as if to catch me lying. Here, Archie is willing to pay Janet and me more money than we've ever made, to do something we've never done before, and he hasn't even asked if we know a tree from a tea cup.

Gingerly, I lift the first log—stick? what are they called?—and check the blue mark. The gloves Archie gave us are too big. My new boots weigh a ton. And then I forget everything, concentrating on getting the logs to go where I want them. One at a time. Carefully. My watch must have stopped, because surely it's more than thirty minutes later when I notice that the back end of the chain—my end—is packed full of wood, three and four pieces deep, while it continues to fall in an alarming blue and yellow torrent at Janet's end. I wave at her, point at the pile. Janet wiggles at her ear with her too-large glove and mouths the word, "Damn." We both bend forward, working harder. But we can't catch up. Hardly any of the logs are going up top, where the machine piles them; they're all getting pushed down here by that little white hand, all rejects. Janet and I keep forgetting which mark goes where. The wood won't lie straight. I get my pieces crossed with Janet's, and once a cart is almost full, it's so hard to get the last pieces to the top. Suddenly, with a long scream, the whole chain grinds to a halt.

In the shocking silence, I catch sight of Janet, the top of her head bizarrely swollen by the red hard hat, bulging bug eyes between the bright yellow ears, her face sweaty and red. I look like that, too, I think, as Tom flies around the corner, frowning. When he sees the chain, he grins. I giggle. Then I'm laughing, or maybe crying. Tears pour down my face, mixed with sweat. We've found the only job for us in the entire north, and we're about to be fired after less than an hour! It's hilarious! I wonder if I'm hysterical.

"Let's get this moving," Tom calls into the strange calm of the mill. He works with astonishing speed to help us pile, and in no time every piece of wood is on its wagon and the wild clanking of the chain starts again.

When the lunch horn sounds, I'm amazed at how tired my arms and legs suddenly feel. We follow the others to a narrow shack in the centre of the mill where we sit on wooden benches at a plywood table worn dark by use and where the walls are papered with posters warning blood if we don't follow safety procedures. Last one into the shack is a short woman with dark hair who introduces herself as Dean, the lumber grader. Dean introduces the others: Cowboy, the forklift operator, who keeps his eyes firmly on the table; Wayne, the man on the chain above us; and a man with bright red hair who introduces himself as Brian.

"I run the front-end loader."

Dean seems to feel Janet and I have imported women's liberation into the north. "Now we'll show these men a thing or two!" she boasts. But I'm afraid I'm going to disappoint her. I feel an unfamiliar, primal urge to guard my energy, waste nothing on conversation. As I walk back to my place on the chain, my arms swing like two sticks from the rigid pole that's all that's left of my shoulders. Four and a half hours to go.

A quarter to five is the moment when I think I'll die. If I don't die, I'll throw up. My arms and legs scream to stop. *Fifteen minutes more,* I promise them. I put my head down and heave, pushing one log at a time, toward five.

Then an angel with red hair appears.

"I've finished loading!" Brian mouths, but I barely glance at him. Seeing he knows how it's done, I only pretend to move my leaden arms while Brian piles the remaining wood.

As the quitting bell rings I say out loud, "I never knew there were so many logs."

"Boards," Brian says. I can hear him. They've turned the chain off. It's ecstasy to stand still.

Every morning for a week, I wake up stiff in a new part of my body, but I love this work. Janet and I get up at 6:45, pull on blue jeans and T-shirts, wash our faces, eat, fill the thermoses we bought at Fields in town, pack lunch and leather gloves into our daypacks, pick up our hard hats and stand out on the highway to hitch to work by eight. Work on the planer chain starts awkwardly every time, but after an hour or so, I get into the rhythm. Turn, bend, pull; turn, bend, pull. If I use the boards' own momentum, I only have to guide them into place, like obedient dogs on a short leash. By week two, my body feels slim and strong, and I hardly ever have to stop to remember what each blue mark means. One morning when I pile two boards at once, it feels so good that I laugh out loud.

We continue to stay with Parminder, who is wonderfully generous. Janet and I chip in for groceries but Parminder insists on doing most of the shopping and all of the cooking, welcoming us home every afternoon from work. She misses the company of the planer mill, she tells us, and is eager to hear news. However, her husband is clearly not as happy with houseguests, so Janet and I start looking for our own place. Accommodation is scarce in the Fort, and eventually it's Brian who tells us a friend of his has just dragged an old cabin out of the woods and will let us live there rent-free for the rest of the summer.

When he drives us over the following Saturday to check it out, it's clear that when he said "cabin," he was speaking loosely. This one has four bare wood walls, a roof and a rough counter inside the back door with a shelf above. That's it. There's no furniture, no drywall, no windows (at least none with glass). No door. From inside, you can see daylight through the knotholes. No light, no heat, no water.

"No problem," says Janet.

Terry, the logger who owns the place and lives just down the ridge with his family, lends us a bucket and walks us to a tiny stream that has drinkable water. Certainly, I have to agree, the price is right and the view spectacular; the cabin is perched on a ridge above the Stuart River valley that winds below us like a diamond necklace through green, with fireweed blooming in every open space like pink lace.

After Terry and Brian leave, Janet and I walk to Fields and buy a Coleman stove, a saucepan, a frying pan, a thermos, a kerosene lamp and fuel, some leather patches for our gloves, and a bag of groceries. We borrow a shovel to dig a hole that will be discreetly hidden by a small bush, and set a roll of blue toilet paper beside it. Janet is matter-of-fact so

I am too. We spread newspapers on the floor to keep out drafts, lay out sleeping bags with Space Blankets on top, and stow our new kitchen gear on the shelf above the counter. We're home. That night, Janet volunteers to cook if I'll fill the kerosene lamp.

"Take it outside," she says, so I carry the tin of kerosene and the shiny red lamp outside and set them on the ground, but in the course of filling the small reservoir at the lamp's base, I splash kerosene over my hands, the lamp and the ground. Quite a bit of kerosene. We should've bought a funnel. The cotton wick waves, a cheery little flag as I fiddle with the length. Guessing, I leave a good inch exposed, then strike a match.

By the time Janet rushes to the back door, I've got the wick burning. Unfortunately, so is the knob at the top of the tank as well as the ground all around, now flickering with patches of blue flame. Actually, the wick is doing a little more than burning—it's a billowing torch.

"Turn it down!"

The lamp settles to a steady burn, though its bright red is sadly blackened, and soon the ground stops burning. I grin at Janet, ridiculously proud of myself—I'm a woodswoman!

Every night, we take turns cooking, and after supper and on weekends, we read or write. I've made a "no sex" rule for the summer, trying to keep my life simple, and I only slip a few times. My favourite place is outside, alone, sitting on the pile of boards under our front window overlooking the Stuart River valley. There's only the occasional sound of kids playing and the stink of mosquito coil—northern incense. I'm reading Kate Millett's *Sexual Politics* and Carlos Castaneda's *Tales of Power*.

Some Fridays at work after the chain closes down, we do a quick cleanup and then meet in the lunch shack where Tom already has two cases of Molson open so we can toast the number of train car loads we've filled that week. Any excuse for a drink.

—

One Saturday, Dean invites Janet and me to her house for a "girls only" party that soon becomes a regular event. After a couple of Saturdays, Janet stops coming because, she says, there's too much alcohol, but I keep going for two reasons. One is that it bugs me that women are so cautious. Why shouldn't we be able to drink like men? The other reason is that I'm intrigued by the talk. I'm twenty-eight and have had many lovers, but even in the women's consciousness-raising group I joined last year in the city, we rarely talked about sex. Most of the women here, in this group in tiny Fort St. James, are ten years older, married, with kids, and no illusions, and nothing is held back, like the time Dean proposes a toast to Nora as the first—and so far only—woman hired in the sawmill that companions our planer mill. Dean raises her rye and ginger.

"To Nora!" she says, "who's got everything it takes. And I mean everything!"

Nora grins and says something about having everything except something good between her legs more than once a year. I'm shocked, but the other women laugh.

"If you want something done right, you have to do it yourself," one of the women says, and I'm still more shocked—but it's delicious.

—

At work, Janet and I have begun to play games to make the time go faster. One morning when there's a rush of boards from Dean's grader shack, she taps me on the shoulder.

"Watch!" she mouths, and sweeps up four at once, delivering them perfectly onto the cart. I try the same. Bull's eye, *thonk*! We practise piling five, six boards at a time and climb them, even when the level of the cart is higher than the chain. We're fast, we know, and graceful. In fact, piling lumber feels like dancing. Why has no one ever told me what a pleasure it is to feel my muscles firm beneath my clothes, to have this endless energy, to feel alive in my whole body and not just from the neck up? It's as if I've been dozing all my life until now. The mosquito and blackfly bites and bruises up and down my legs hardly register. When I confess to Dean how much I like this work, she tells me I'm not the only one.

"Archie's smart. Nobody else'll hire us, but he figured out real quick we women are so damned grateful for the job that we turn up on time every day, never drunk or hungover, and work our butts off. We're thrilled to make this kind of money even though the men make more. And if there's an equipment breakdown, we grab a broom and start sweeping instead of heading out for a smoke."

"What do you mean, the men make more?"

"Every guy pushing a button here makes at least a nickel an hour more than you," Dean says and laughs, though the joke's on us.

One Saturday night, we're all invited to a tree planters' party. All I've seen of planters is when they're in town doing laundry or collecting mail, short glimpses of people who move like gypsies in long skirts, bright colours and brightly patched jeans. Their hair is long, too, loosely tied with leather or a tattered ribbon, and they smell of something exotic that someone says is patchouli. They seem neither entirely female nor entirely male, and they laugh a lot.

Janet refuses to come because of the inevitable drinking but I go with Tom, Brian and another man from work, all crammed into Tom's four-wheel drive, bumping and rolling over rocks and stumps into heavy bush. I've never felt so far away from everywhere. The camp is a cluster of white tents thrown like pearls over a dark velvet clearing on the edge of a small, round lake. Brian mixes me a mostly-vodka-token-splash-of-tomato-juice-in-a-beer-bottle concoction and we sit down to share a joint with a tree planter—but I'm restless. I wander off toward a rough opening in the forest and once over the first ridge, still sipping at my drink, I'm out of sight of camp. I follow the road, paved with tree stumps a foot high, as it leads toward a huge, low-rising full moon. When I raise my hand to drink again, my skin shimmers.

I've learned men's way of working and now men's way of drinking. My whole life, I've been taught to pay attention to what men want, to please them above all. What's becoming clear is that I don't know as much about women, and somehow, I'm beginning to learn it here. The farther I walk, the more relief I feel—joy, even. This walk through trees feels like a ritual, a wedding aisle, with me, the silver bride. I don't know who I'm marrying, but trees are my witness.

Back at the camp, someone has started a huge fire and several half-naked men and women are dancing around it to the beat of drums. I'm drunker than I've ever been, but I feel anonymous and safe, so when I notice a small wooden sauna near the water, it seems natural to take off my clothes and leave them beside others on a bench by the door.

The sauna is lit only by the light that escapes through chinks in an old wood stove burning in one corner, and the air is a bath of heat and marijuana and patchouli oil. The shadows of four or five men flicker on the benches in front of me and I'm already sweating as I climb to the middle bench. The men occasionally talk quietly, laughing and smoking dope. There's something wonderful about sitting naked and silent with strangers. This is how a goddess must feel—so confident that she can show her breasts and her muscled shoulders and thighs without fear. When I leave the sauna there's cool shock as I step into the lake, soft mud squishing between my toes like Jell-O before the bottom firms up and I swim lazily to the centre, utterly at ease.

The following week at the mill, Tom tells us that on the other shift, Wayne—trying to straighten out boards on the automatic piler above us—caught his hand under a two-by-four just as the sledgehammer came down on it. On the drive home, Brian explains matter-of-factly that Wayne did it on purpose because he hated his job.

I'm horrified. How could someone deliberately hurt themselves? Why didn't he just quit?

"No compo." Compo, he has to explain, is Workers' Compensation. Because Wayne was injured at work, he'll get paid until he's better.

It's the first time I see the chasm between my pleasure in this work and how differently others might feel.

Now the guy who used to be strapper—who tied metal bands around Janet's and my loads for loading into the rail cars—takes Wayne's place, and Archie hires a tall, well-built guy who strapped at a mill in Smithers as replacement. But when I push out the first cart, it's clear from the way he pokes at the strapping machine, the new guy has no clue.

I walk over and, trying not to hurt his feelings, say, "Is this a new kind of strapper for you?" And show him how.

That night on the way home, I fume. "He's never used a strapper in his life! He's being paid more than Janet and me to do a job he doesn't know how to do!"

Brian says calmly, "He got it done."

"I did it for him! He lied!" I'm boiling with righteous indignation but Brian just shrugs.

Toward the middle of July, Archie hires a new labourer, Brett. Whenever Brett sweeps around our area I notice a wonderful smell—sweet and fresh like spring but richer. Irresistible.

When I ask him, casually, what aftershave he uses, he looks surprised. "Soap."

He must be joking. "No, really?"

"Really." He has beautiful eyes.

Janet doesn't even notice him, but I can smell Brett twenty feet away over the competition of pitch and sawdust. I begin to follow him around by the nose, so to speak, and after a few parties, my no-sex rule goes out the window.

Sex with Brett is easy—the rest of my body following my nose into the sweetness of him. He is tender and not what you'd call goal-oriented. One night as we lie in bed at his place, he tells me he's on parole.

"Parole?" I pull away to stare at him. "From jail?"

He laughs a forced little laugh. "I was stupid." He recently served time as a drug dealer, he says, which is why he's here, on parole, living with his mother. He is—was—an addict himself.

Do I like him less?

I wish.

—

Though I love lumber piling, it's starting to get repetitive, and one Friday Brian asks if I'd like to learn his machine, the tilt-hoist.

Sitting like a fat frog at the mouth of the mill, the tilt-hoist is a huge metal table that the operator raises and lowers to catch the rough-sawn boards dumped on it by the outdoor fork-lift. I know this machine is important—it sets the pace for the mill—but it doesn't look all that hard to operate.

"I'll come get you at the end of shift," Brian promises. "Janet will cover for you."

But he doesn't come, and that night as he drives us home, he explains, "Archie doesn't think it's a good idea."

So the next day I go into the office before work and ask Archie myself if I can try the tilt-hoist.

"No."

"Why not?"

"Too hard." He doesn't even look up.

I stand for a minute, then turn and head back to work. Maybe it really is too hard. Maybe Archie and Brian know something I don't.

—

Janet leaves first. From the beginning, she'd been clear she was going back to school in September to finish her teaching degree. I consider staying, but within days of her leaving there's frost on my space blanket every morning. I've loved it here for the summer, but I don't want to be a lumber piler all my life. I've been officially accepted for a master's degree in the Communication Department at Simon Fraser University and now I have the money to pay for it. I give Archie my notice, Brett and I promise to visit each other, and I head back to the city. The academic life isn't exactly what I want, but what else can I do?

2

madness and metaphor

After the casual lifestyle of the Fort, where everyone wore T-shirts and jeans and there was almost nothing to buy, Vancouver with its billboards and ubiquitous ads is shocking. I find a communal house to share with three other students, and I like it that we agree to divide all expenses—until the other woman questions why the men's food receipts are just numbers scribbled on bits of paper.

There are no receipts, they confess, because as an act of defiance against the capitalist system, they steal the food. But it's fair for us to share the cost because... and the argument loses me here. Something about them taking the risks of stealing.

"So don't steal!"

"It's a political act!"

I move in with other friends from school, Linda and Paul, who live in a communal house as a sort-of-couple, mostly for the convenience of looking after young children whose care they share with partners elsewhere.

—

I was used to having kids around. I'd grown up in Montreal, the eldest of six. My job at home was always to "look after the kids," and by the time I was twelve I was clear: I was never going to have children. I'd already had them. So all through my teenage years, necking with boys in the back seats of cars, I pulled back whenever a hand reached for my breast. Everyone knew about bad girls who went all the way and had to suddenly disappear.

Then came the Pill. I was twenty before I heard of it. Now we could have all the sex we wanted and never get pregnant, and the doctors assured us there were no side effects. I did the deed—getting rid of my virginity—with the office clerk on the yellow shag carpet of my first apartment, as Simon and Garfunkel crooned from the hi-fi. Afterward I thought that if that was all there was to sex, it was no big deal, though the more I did it, the better it got.

If you were a girl who needed a career in the 1960s because you were never going to get married, you had three choices: you could be a nurse, a secretary or a teacher. Teacher felt too much like parent to me, and I wanted nothing to do with nursing, so—as the first in my family to go to university—I picked a small school, Mount Allison, that not only offered a BA degree with Secretarial Certificate, but was in New Brunswick, twelve hundred miles from home.

Distance was important because my dad drank. There were some sweet moments—Saturday food shopping trips with Dad when he'd buy us all the pop and chips Mum never would, and Saturday night parties with friends and neighbours gathered in our TV room, the furniture pushed aside for dancing, the hi-fi turned up so we kids could fall asleep to the music of Tommy Dorsey, Duke Ellington and Satchmo, as Dad called Louis Armstrong, his favourite. Later, there'd be Dad drumming on a pot and the waggly voices of adults singing "Down by the Old Mill Stream" and "Daisy" along with Mitch Miller. But mostly—increasingly, as I got older and Dad drank more—I thought I would die if I stayed there. There was no other word to describe the feeling of unpredictable, omnipresent danger in our house. Every weekday I waited for Dad to come home, keeping my breathing shallow so I could hear the instant the front door opened, listening for how he would be this time, for the clink of ice.

Most evenings, Dad worked late. But on Sundays he insisted the family eat together, the traditional Sunday "din" of roast beef, Yorkshire pudding and apple pie he remembered from his boyhood in England. Once I got to be twelve or so, I began to notice how, as soon as we sat down together, Dad would start in criticizing Mum—her looks, her clothes, her attitude. "Farm girl," he'd say, and take another drink.

No one talked back, no one challenged him, and finally I couldn't bear it. My mother had always told us, "You have a mouth, use it," so I did. I defended her. After that, Dad's ridicule shifted to settle on what he called "the boring masses," people who took the eight o'clock commuter train to work every morning and back again at five, who carried brown paper lunch bags and lived, he said, monotonous lives. I defended them, too. Why belittle someone just because they didn't get paid enough to afford lunch out, a drink or two, like Dad? Our Sunday night arguments became a family tradition, then a family joke, but I never found them funny. Soon, I was always angry at him—at what I saw as his smooth sense of privilege, his outrageous confidence that if his charm couldn't get him what he wanted, then his growing salary and prestige as a rising star in his textile company would. All I could do was poke tiny holes in his complacency.

I began to ask questions. When I complained to Mum that Dad always got his way, she said, "Someone has to give in."

"But why does it always have to be you?"

She didn't answer.

Always, my escape—my refuge—was in reading. I read everything and anything I could get my hands on, at every spare minute. Mum bought *The Book of Knowledge* to try to keep me supplied, pointed out the local Bookmobile, and eventually subscribed to the Reader's Digest condensed books. I read them all, cover to cover. In books, I was in another world, safe.

At age twenty-one, after a post-graduation trip hitchhiking around Europe, I came back to Montreal to look for a job—as a secretary, of course. But it took a while, perhaps because my heart wasn't in it. One day when I dragged myself home and used the word "boring" yet again in front of my mother, she said, "If you don't want to be a secretary, then what do you want to be?"

It popped out before I could think, surprising me. "I want to be a writer."

"Then get a job as a writer."

Sure enough, that night in the newspaper I found an ad for a job in the public relations department of a large multi-national corporation: assistant editor of their house organ.

I loved the writing, but when I found out our company manufactured ammunition and explosives, some of which were going straight to Vietnam, I had no words for my uneasy feeling. Between that and an unhappy love affair, I quit and bought a one-way train ticket across Canada. I got off and stayed at government-run youth hostels along the way—Parry

Sound, Winnipeg, Calgary, Banff. Sometimes I even applied for jobs but I never went back after the first phone call. I climbed back on the train until I reached the end of the line in Vancouver.

The guys at the Cool Aid hostel said I had to see Stanley Park so I hitched downtown, walked up a low bank—and stopped. Towering over me were trees like none I'd ever seen. Their trunks were vast, and it made my neck hurt to look up to their tops. The air was heavy with the smell of evergreen and salt and just like that, I knew I was home.

—

I got a job at a museum arts store where one of the customers was a bookstore owner named Van. Within a few months, we'd moved in together. In 1971, that was a new option. Apart from the fact that no one I knew had actually done it and that there were no instructions, moving in together was easy. I wrote my parents a brief letter telling them the news, and a week later got a carefully worded reply from Mum saying she and Dad hoped I'd be happy and would one day want to formalize my relationship with Van by getting married. Clearly she and Dad were horrified, but they were three thousand miles away.

I'd spent the last few years of my life as a single woman happily making my own choices, but oddly, as soon as I moved in with a man, whatever he wanted was fine by me. Van called me his little girl, and I fussed over him. He was seven years older, and, I was certain, much smarter than me. I read what he read—Beckett, Krishnamurti, the *Tao Te Ching*, Joseph Campbell and the *Bhagavad Gita*. When he wanted to try Transcendental Meditation, we both took the class, though I continued to meditate long after he lost interest. He loved classical music, and though I'd always found it boring, now, to please him, I curled up with him in the yellow easy chair to listen until after a while I began to like it too. Sort of.

One of Van's friends was a lanky, sixtyish man with a ring of long white hair around a balding pate, who taught at the newly opened Simon Fraser University. Fred was full of ideas and widely read. He dropped by our house often to talk.

After the art store, I'd worked at a series of government-sponsored jobs, but I was restless, and Van and I talked about me going back to school, maybe getting a master's degree. In 1973, after a summer of deadly boredom typing library cards on a manual typewriter for the school board, he agreed to pay for my tuition, and I signed up at Simon Fraser University to see what—if anything—I might be interested in.

At the registration desk, when they asked my name, I hesitated. I'd always been called Kathy.

"Kate," I said, and felt instantly tougher.

—

I'd never met an intellect like Van's or Fred's, so hungry for ideas they sat up until two and three in the morning, drinking coffee, smoking dope and arguing passionately. Often I couldn't follow all the details but I loved the clash and battle of the conversation. We were all agreed—we were talking about how to change the world. When Fred asserted that our society had outgrown the old-fashioned nuclear family and needed "broader bases of connection," different ways to live together, I felt a rush of recognition.

I'd felt this strongly once before, five years earlier. It was on one of the rare days when I had my parents' house to myself. It was a burning Montreal afternoon, and in the dim cool of the TV room, I sat reading a thick blue paperback I'd just bought, called *The Feminine Mystique.* I was in my third year of university and I'd never before read a book that was just about women. The author, Betty Friedan, had identified what she called "the problem with no name": women who had everything they could possibly desire—a loving husband with good prospects, beautiful children, a house with an electric fridge and a fancy new vacuum cleaner—but still weren't happy, women who felt empty even when their husbands took them for nice vacations, who felt "trapped."

I was electrified. I'd always asked questions, always had this restless curiosity, but usually the questions were so vague, I barely had a name for them, and I seemed to be the only one asking. Now Friedan was asking the same questions I'd been asking at home, at university and at church.

In the Baptist and then the United churches that our family attended, I'd always loved the drama of Bible stories, the struggle and suffering of the saints, even if they did mostly belong to the lucky Catholics. I loved the rituals, too, the sudden hush when we walked into the sanctuary, the sense that there was something bigger even than my parents. In Sunday school I was a star at memorizing Bible verses. I did it for the fun of winning more gold Keys to the Kingdom to add to my orange plastic key chain, as well as for the roll and heft of King James's verses on my tongue. "A good girl," people said, and I preened. By age thirteen I was teaching Sunday school, singing with Mum in the adult choir, and planning to be a missionary when I grew up.

But I had questions. Why did girls have to wear hats, but not boys? Why did girls have to stay clean while it was okay for boys to get dirty? Why was getting kicked out of Eden all Eve's fault?

My mother said, "Ask the minister," but the minister was no help.

"Have faith," he counselled over and over. "Just have faith."

In what? When I joined the Unitarian youth group so I could spend my Saturday nights around a guy named Mike, the minister in our church denounced from the pulpit any "so-called church" that allowed young people to dance in their sanctuary. After that I still went to my own church, but my heart wasn't in it. I wouldn't be a missionary after all. Then I stopped going altogether.

At university I'd vaguely wondered why it was assumed I'd get married, why my professors were almost all men (except for the typing teacher), why it was mostly men who spoke in class. I'd written in my journal, "There's a forever question mark in my mind—sometimes I can't even put a name to it." That was it; mostly I couldn't even put a name to my questions. Until now. Now, Betty Friedan spoke of women's "quiet desperation," of being stopped in our development far short of our fullest human capacities. She not only asked the questions, she offered explanations that—for the first time in my life—made sense. Men's dominance in society, she wrote, was based on physical force and economics, and it wasn't fair.

I laid the blue book down on my lap and I knew my life was changed. I wasn't crazy; I was a feminist. A short time later when I read Mary Daly's article "God is a Verb" in the new feminist magazine, *Ms.*, my questions would intensify. Why does the church call women's bodies "unclean"? And if women's, why not men's? And whoever heard of a male "creator"?

Which did nothing to improve my relationship with my father. Now that I had a name for his arrogant manner—male chauvinist—I was more angry with him than ever.

—

Since Fred was spending most nights at our house anyway, it seemed natural that when Van and I moved closer to the university, Fred moved in with us. Soon after, he started bringing an old friend, Susan, into our nightly conversations. Different people came and went, but "The Group," as we called it, settled down to be Fred and Susan, Van and me, and a grad student named Glen whom I'd met in Fred's class.

Apart from a comment now and then, my main contribution to our discussions was a steady flow of tea and coffee, though occasionally I ventured that some of our theory didn't feel entirely, intuitively, right.

"Intuition is the sum of *past* experience," Fred assured me. "It can't be trusted as a guide, because it points backward."

We were part of history, of a revolution that was also taking place in France and Germany and Spain. We'd clean up the mess our parents had left and build a brighter world starting here, starting now. We were creating a new community, an intentional one planned for human happiness. Of course, this meant practical changes like moving to the country, like experimenting with traditional, closed couples. In the spring of 1974, The Group bought a large piece of property north of Vancouver, and Van took a lover. We discussed it calmly in advance, and on the evening he first slept with her, Fred and Susan and Glen stayed with me, sharing the anguish.

"These are the old structures speaking," Fred said of my pain.

I wrote it down in my journal. The small certainty of black lines on a page made change—necessary as it was—feel a little more concrete, a little less scary.

I was surrounded by powerful, articulate men, but for some reason it emphasized the importance of affirming my female self. At school I joined the Women's Caucus and went to women's meetings, devoured writers I'd never heard of before: Virginia Woolf, Kate Millett, Germaine Greer. I began to see how the public sphere was dominated by men and their overly rational way of seeing things. I was intrigued by different concepts of gender. What was it to be female? How was "female" different from "feminine"? At home, when I brought up the question of the equality of women, the men briefly nodded as we passed on to something more important.

Glen and I became lovers. It seemed a natural thing to do, given what good friends Glen and I were, and The Group's theory that we needed to break from traditional couples. So I was surprised when Van reacted badly. More than badly. He felt betrayed, he said. Humiliated. Even Fred seemed to have a hard time with me having a lover. "Don't you still love Van?" he asked. I was confused. How could this experiment not be the same for both of us?

When Van threatened not to pay my tuition for the next term, for the first time in this experiment I got mad. I moved into Susan's house. The separation made me crazy, but wasn't this what I meant when I said I wanted to be an independent woman? When Van held true to his threat not to support me any longer, I got a part-time job, then took a two-week babysitting job looking after two little girls while their parents were away.

On the first Sunday evening, I'd just put the kids to bed when Van phoned to ask if he and Fred and Susan could come over to talk. When we sat down around the kitchen table, the three of them were oddly quiet. They'd had a bad acid trip, they explained, as I poured tea. Fred lit a joint, and Van opened the conversation with, "We've made a mistake."

Monogamy, they'd decided while on acid, was the only possible relationship after all. Now Van wanted a decision: either I agreed to stop seeing other men, leave school and commit myself to him alone, or he never wanted to see me again.

I was stunned. Hadn't we agreed to broaden our community by not tying ourselves to a single other, agreed that these old emotions were merely the result of conditioning? I heard Van say, "drop out of school," I heard "monogamous"—but for me, not for him.

"A man looks for the *form* of a relationship," he said.

And they left me to think about it. But I wouldn't abandon what we'd worked so hard toward. I found a communal house and moved in with three other students while I kept dating Glen and others. Glen didn't like it but I was clear with every man: I dated others, I slept with them. It was my commitment to experiment. Besides, I liked having lovers, plural.

My women's consciousness-raising group was my anchor. No matter how crazy or lost I felt, when I heard other women talk about their feelings, I was deeply comforted and reassured. When my friend Sandra's boyfriend broke up with her, telling her she'd been no more to him than an "intriguing fly in a jar," we summoned our feminist best friends to a wake at Hy's Steak House and talked about how even the men who said they loved us feared us. What irony: they were afraid of our strength, and so were we.

Linda and I started a column on women's issues for the campus newspaper, calling it "Surfacing" after Margaret Atwood's latest book. I joined the Chile Solidarity Committee, but it was mostly the men who talked. It took courage for us women to speak up. Using the word "sexist" was like pulling out some naked part of our female bodies. If one of us finally got brave enough to demand a response, the men reassured us by saying the Woman Question would be dealt with *after* the revolution.

Soon I was going only to women's groups.

At school, I was taking four graduate courses plus doing a research job, working part-time in the library and as a teaching assistant. Exhaustion was a friend.

Work was one anchor; sex, another. I couldn't keep track of who was sleeping with whom in our communal house so I moved again, into a basement apartment with Sandra. But lovemaking was in the air; it was just what you did. We made love on beds and on desktops and pressed against kitchen counters. We made love drunk, love stoned, love straight. We made love-with-not-a-shred-of-love-in-it, falling-asleep-love and love-bright-and-early, quickies and day-longs. I made a point of telling every man I slept with that I slept with others, too. So why, after a few times together, did the men begin to get jealous? Was I really as hard-hearted as they said?

One night Glen and I went to his parents' summer cottage, partly so we could attend a Keith Jarrett concert nearby, partly to try LSD. We set out sleeping bags and snacks in the cottage before each chewing a tiny square of blotting paper.

The concert was in a pleasant wooden hall. As we sat down Glen brushed against my arm, and my skin fizzed as if I were bathed in mineral water. I didn't know jazz but I loved the mental journeys Jarrett's music took me on, his intensity, his sudden out-of-tune humming as if the music had taken him over. When we left the concert it was raining, and the wet felt sweet on my skin. At the cottage, Glen retreated with a stack of old magazines while I sank into an overstuffed chair. After a few minutes I noticed the sound of a fan clanking, like a cheap guitar stuck on one note.

A man built that fan.

I was surprised at the thought. It sounded a little hostile. I liked men. *And a man installed it.*

Had I say it out loud? Was this what they called freaking out? In the far corner, Glen pored over his magazines.

"Do you hear that fan?"

He looked up, listened. "Yes." Went back to the magazines.

A man built this house, I thought to myself. Suddenly I couldn't wait to get outside, to where the moon had filled the potholes with silver. I walked to a tree and sat with my back to it, breathing pine and wet earth.

A man built the road to this cottage. A man drove the grader. Men mined the gravel and processed the tar and built the machines that built the road. It went on and on in a demented version of the House That Jack Built, only this was the entire world—all built, all run, by men. No women.

Back at the apartment I shared with Sandra, I started having dreams about an old woman. I knew her as a gypsy by the bright colours she wore, by her long skirt and the ribbons and scarf around her hair. She was small and bent, with wise eyes. She watched me intently in dream after dream as if she knew something, but she never spoke.

—

I'm a boy living on the top floor of an old hotel that overlooks a valley. Time bends in this place; we receive a letter written on the hotel's original antique vellum stationery, a letter from the past.

One day when the old gypsy woman comes to visit, she takes me and another person for a drive. She's done this before, but this time she's in a hurry. As she hitches up the horses, she glances back over her right shoulder to be sure we're still here, and then we're off, the gypsy driving, sitting in front on a bare wooden seat. The ride is oddly silent. There's no jingle of harness, no birdsong, no wind, no one speaks.

Always before on this road, after we've crossed the bridge into the valley, we've turned right, but this time the old woman turns left. I start to tell her this is wrong, but as she turns her head over her right shoulder to listen, her face begins to change, to melt and become younger. The landscape too, is changing, evaporating, and I understand that we're being carried into the past.

When I wake from this dream, I am not afraid, but I know I'm about to go somewhere unexpected.

—

Now, on a clear April evening in 1976, I hand my ticket to the Greyhound driver and climb on board the bus for Prince George. I'm going for another visit with my friend Brett, who's moved from Fort St. James. I go for his intoxicating smell, for his humour, for the companionship and good sex and perhaps even for the danger. The last time I came north, we'd ended up in a small cabin in the woods with three other ex-cons, pressing a knife blade over a lump of hashish on a hot electric burner. The more stoned I got, the more alert and cautious I became, as if looking through ice.

On that visit, Brett gave me the book that now lies on my lap, *The*

Day on Fire, about the French poet Arthur Rimbaud. "Drunkard, dope user, sexual pervert, blasphemer, social rebel," the cover announces. "Debauchery was a religion with him." As we roll out of the city I read, occasionally glancing out the window at the rising moon.

Just beyond the town of Hope, I get nervous. More than nervous—terrified. As certainly as if what I'm reading is a script, I realize this book is a message. Brett pretends to love me, but really, he hates me and is going to kill me. Part of me knows this is madness. I take deep breaths—be reasonable! But the stories in my head circle and all come back to this: if I go to Prince George, I will die.

With my legs buckling, hands clammy, I work my way to the front of the sleeping bus and tell the driver I need to get off at the next stop. He looks at me oddly, but I don't care. Cache Creek? Fine.

There's no bus back to Vancouver tonight, he says, but *cacher* means "to hide." I'll find a cheap motel, and tomorrow morning I'll take the bus home. I feel a surge of love for my seat mate, who snorts in his sleep as I step over his legs and sit down again in the window seat, clutching my coat tighter around my shoulders.

Twenty minutes later the bus pulls up at a small building where RESTAURANT flashes off and on in pink neon. I use the pay phone outside to call Brett, tell him I can't come after all. I hang up before he can ask why. Then I dial Linda in Vancouver, but she's not home. Instead, her new roommate, Joyce, answers. I have to tell someone, so, over the sound of my heart pounding, I tell Joyce, a stranger, that I'm going mad.

Joyce says, "Here is an opportunity."

She's crazy too.

"An opportunity," she repeats, and then I remember she's a therapist. She knows crazy people; she's heard us before. I press the receiver tight to my ear. A thin line.

"You've been sweeping things under the rug for years," Joyce continues, "but the rug can't cover it all up anymore. This is your chance to clean things out, start fresh."

"Thank you," I say. I hang up and walk into the restaurant, where I order tea with milk, holding the steaming cup with both hands.

When the bus leaves with all its passengers, I'm alone in the dingy restaurant except for an older couple sitting at the next table. The woman glances at me, curious.

"Off to Vancouver?" she asks.

"Yes. But there's no bus until tomorrow."

After a glance at the man, she says, "We can give you a ride."

So by midnight, oh-thank-you-goddess, I am in my own bed with the front door and all the windows locked tight and a note left on the counter for Sandra telling her I'm home.

—

For the next few weeks I don't leave the apartment. After my first brief explanation, Sandra asks no questions. I lie on my bed and shake, certain I'm in mortal danger but not sure from what or who. I write in my journal, "I'm a Good Girl waiting for the prince—oh please! please!—who will save me." Everywhere, I see hypocrisy and contradiction. All I really—desperately—want is to die the small death of orgasm and in that way lose myself. I look back over my long line of recent lovers, and it's clear I'm as guilty as Brett, though my drug is sex.

When Glen comes to visit, he plays me the new Eagles' song. The chorus croons, "Take it easy. Don't let the sound of your own wheels drive you crazy," but I can't handle even comfort, and send him away.

When I venture out in public, the only place I feel safe is in the park, surrounded by trees, water, birds. It's the physical world I trust. I avoid dope, devour books about crazy women that I bring home from the library: *You Never Promised Me a Rose Garden, Women and Madness, Sybil* and a biography of the poet Anne Sexton, who committed suicide. Women, even women who write about being crazy, make me feel sane, as if madness is the only acceptable way for a woman to show rage. If I can't win your respect, we mad women say, I'll win your horror.

When Brett gets caught dealing drugs again, I visit him in jail in Vancouver, and seeing him behind a glass shield breaks the spell. Is loving hopeless men my last barrier against giving the best of my energy and attention to myself? Desperate for answers, I find a therapist.

Therapy is like conscious dreaming. The spotlight it shines on my own history? Pitiless. The therapist says, "What you try to forget doesn't die. It only gets buried alive and will come back to haunt you." My demons struggle to stay hidden, but I feel such relief at not having to continue the hard work of keeping them buried. I loved Brett because he was a rebel, and I start to understand. Me too. Underneath, we felt like losers.

In therapy, I begin to see metaphor, and that metaphor matters. I dream that I'm trying hard to please people by wearing a beautiful long silk dress, but when I turn around, my blue panties show through and I think, "I'll have to change." The woman beside me, who has braided hair, begins to choke because her braid is too tight.

When I write down such dreams, the messages are so obvious they make me laugh. I'll have to change, and "braid"—my name, my identity—is choking me. Letting go of literalness is a liberation.

At school, I'm learning something else that feels like a key to my sanity: every choice doesn't have to be "either/or." Maybe my only options aren't either total submission or gut-stubborn aloneness. Maybe there's a third way.

The person I spend the most time with is Tim, another graduate student. I love his intellect, and our lovemaking is searingly intense. We often study together, and when we take breaks, he plays his guitar, even shows me how to play a few chords. He wants me to come live with him in a commune in Maine. But every time we discuss it, my fear of dying comes back, powerful as ever. By now I understand that it's a metaphor, that what I'm afraid of is losing control, being hurt—dying emotionally. Every change, I remind myself over and over, means a necessary death of the old way of doing things. Marx called it thesis, antithesis and synthesis, and applied it to all of history.

Maine becomes all I think about—to go or not?

I need to talk to Van. I go to visit, and by the end of our conversation, I feel crazy again. When I tell him I have to leave to go to an appointment with my therapist, he asks if he can come along. For the next hour, the therapist watches, listens, says hardly a word as Van and I talk about The Group, about his ultimatum. Each time he begins his rhetoric, I cut through it. Even when he calls me a whore, I'm not pushed off my centre.

"I was consistent," I tell him. "I was loyal to what we all agreed on. I didn't object to your having a lover, so why was my having one any different?"

In school, another thing we're studying is double binds, how saying one thing while doing another is crazy-making. When the therapist nods at the clock, there's something I have just come to understand.

I would question everything, I say, more to myself than to Van. I need to hear this out loud. "I'd question everything, including my most intimate relationship, before I would sacrifice truth." I feel saner than I've felt in years.

⟶

There's only one thing I'm certain of: I must get out of the city. It's August, the end of summer term. I decide I'll take the fall term off school to get my bearings, come back in the spring. I ask around, casually—does anyone know a place I can rent, cheap, for three months? A lot of young people are moving to the Kootenays in southeastern BC. It's quiet, they say, beautiful in the mountains.

"I don't know about the Kootenays," a friend says, "but I know of a cabin on the Gulf Islands."

By coincidence, I know the island she names—Pender Island. A friend was married there a few years earlier, and Van and I had gone over for the ceremony. On impulse I call and arrange for the real estate agent to meet me on the island that Sunday, so I can "take a look." I don't tell anyone where I'm going.

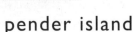

3

pender island

The ferry from the Lower Mainland of British Columbia travels across the Gulf of Georgia through a series of small, wooded islands scattered like pebbles across the water. Tourists take the ride for the beauty of it, but on this day, though it's sunny, I huddle inside, burying my head in a newspaper the whole way across.

The real estate agent is a grey-haired woman in a blue suit and shockingly white running shoes.

"Mrs. Kilgour," she says, extending one hand and pointing with the other toward a blue Mazda. For the first several minutes, as we drive at alarming speed along a meandering dirt road closed in by huge evergreens, Mrs. Kilgour concentrates on ferreting out my life history, with a focus on the probability of drunken parties. I assure her I lead a quiet life, but it's all a game; I don't plan to stay. Mrs. Kilgour brings the car to a halt just before we hit a dead end. There's something different about the air here, lush with vegetation and a hint of salt. I follow her along a path that winds between evergreens that smell like summer camp, across a sunny clearing and through a small ravine cool with salal and ferns and tiny pink flowers. We pass an unpainted outhouse, climb eight rickety wooden stairs, and we're there, breathless. Not from the walk, though Mrs. Kilgour's blue bosom heaves. We're standing on a wide, sun-bleached deck in front of an A-frame cabin balanced on the edge of a cliff. The deck juts over the sea, and the huge blue of sky around us is punctuated to my left by the slash of a dead tree whose branches end at eye level. A large bald eagle sits there, staring at me.

Mrs. Kilgour had said on the phone that the cabin was furnished. A frayed brown couch sags behind the door, and in front of the window stands a round oak table and two wooden chairs. A narrow room on the right has two plywood counters, one of them holding a stainless steel sink. When I peer down the drain hole, I can see the dirt beneath the cabin. Beside this "kitchen," a well-used easy chair squats beside an odd square of gravel on the floor. There's a hole in the roof. "For the wood stove," Mrs. Kilgour says. "You have to supply your own."

In a second tiny room is a fridge. At the top of the steep ladder to the loft, a green foamie lies on the floor in front of the window that continues up from below. A sailboat drifts past the island in front of me, and beyond that are the mountains of Vancouver Island. Each room has a single bare light bulb, dangling.

Mrs. Kilgour eases into one of the chairs at the oak table and takes a sheaf of papers out of her handbag.

"Rent is $400 for the year."

"But I only need it for three months."

"Take it or leave it."

Once in a while you've got to believe in the dice. "I'll take it." I try to be businesslike as we finish up, but I'm fighting a strange desire to push Mrs. Kilgour out the door so I can be here alone. When I hand her a cheque for $400, I have enough money left to live here—frugally—until December. The cabin has no heat, no insulation and no plumbing, but there's electricity and an outhouse, and there's even running water—at a tap near the road. Mrs. Kilgour drives me back to the ferry and I spend the return ride planning what to tell Tim.

—

Dad always travelled a lot for his work, and whenever he came to Vancouver on business, he'd invite me out for dinner. We always ended these evenings with another fierce argument, usually about women's rights and working people. Toward the end of every evening, after several drinks, he'd ask if I needed anything and offer to buy it—trying to prove what a great guy he was, I figured. I'd end up yelling, "You can't buy me!" And once, "Fuck off!" which in my family was unspeakable sacrilege. Each time, I vowed never to see him again.

But when I mention in a letter home that there's no stove in my new place, my parents offer to send money for a hot plate and I accept. I hadn't thought about cooking when I rented the cabin, but the bulky white hot plate is my most precious possession when, a few weeks later, Tim and I load all my belongings into a borrowed station wagon. We also make room for his guitar; there are few times we've been together that haven't ended with singing. My horoscope for the day says, "A good day for real estate agreements," but for tomorrow, "Be cautious." Fine. It's today I need to get through. The weather is another good omen. It starts with thick rain clouds, but the closer we get to the ferry, the sunnier it gets.

Part of my thin budget is for a wood stove, but neither store on the island has one in stock and Dave, the man in what they call The General Store, promises to order one. I buy a bag of groceries and arrange to rent a mailbox—number 49. A magic number, I decide—another good omen for me, though Tim is clearly not happy, omens or not. He still badly wants us to go to Maine together.

"This is only for one term," I promise, and tell him he can visit. We sing a farewell song ("Ripple in still water, when there is no pebble tossed"), then he gets up to leave.

"Your guitar," I remind him.

"I thought it might be good company for you," he says. "And I have my banjo."

It's a breathtakingly generous gift, though a tiny voice whispers that he expects to end up living with me soon anyway, reclaiming it.

Finally, the noise of his car fades away and I'm alone, on the deck, looking across water toward mountains. I've never lived alone, never been alone for more than moments at a time. Now I crave silence. I'm fierce for it. I don't want anyone's company. I don't want a radio or a television or a newspaper or a book. Not even a guitar. I want to stand on this deck—my deck—alone. For one clear moment, I feel blessed.

In front of me yawns emptiness. The air brushing my cheek isn't city air. This air has wild animals in it. I shiver, though the afternoon is warm. I'm safe here. Aren't I? I can see for miles. But...

A bird squawks, and I whirl.

Get a grip! I can do anything I want now—anything.

But I have no idea what that might be, so I do something I've never done before: I do nothing. I wait. I don't know for what. When a tiny voice says, "Pick up the broom," I pick up the broom propped inside the front door, and when the voice says, "Sweep," I sweep the cabin. When that's done, I try to sit down again, but instantly jerk back to my feet. Too scary. I lay down the brown raffia rugs a friend donated, make the bed, cover the kitchen counters with a pretty blue and white oil cloth I bought especially for this, and set out the canisters one of my sisters made me last Christmas, pasting pictures on empty coffee cans.

The cabin is the shape of its name, A-frame, built as if someone had leaned two long slabs of wood against each other, the top floor forming the crossbar of the A. Each open end is mostly glass. It's spooky to be exposed behind so much window, but I can't afford curtains. I pick up the two plastic containers I bought at Army & Navy and head back along the path. Water from the hose gushes fast and clear, and as I walk back to the cabin to fix my first supper alone, I feel a reassuring slug-slug-slug with every step.

At nine o'clock that evening it's still light, but I lock the door and go to bed. It isn't until I lie down that I notice the whistle. I stiffen, count: eight, nine, ten. A pause, and again. Every ten seconds from out on the water comes a high sound not quite human. Someone communicating? A countdown? But nothing happens, and after getting up twice to make sure the door is locked, I sleep.

Day one. A-frame. Alpha. Beginning.

The next morning after breakfast, I sit at the picnic table on the deck to read. The eagle is back, but this space feels too big. I can't stop listening, as if something alarming is about to happen. But there's only the occasional high-pitched screech of the eagle and the whistle out on the water that I eventually figure out is a navigation buoy. What I can't get over is the sense of a physical force around me. Vast. Alive. Is this what people call God? Nature? Locking the door won't keep it out, so I wait for it to take human shape. Every time I hear a noise—the creak of a tree, the crack of a branch—I'm sure it's something or someone terrible.

In desperation, I get out the guitar and learn a new chord, E minor. It's perfect. Making noise—making an E minor noise—embraces fear, and for the next few days I spend most of my time outside sitting at, or preferably on top of, the silver-grey picnic table, singing, especially as dusk creeps closer: Bob Dylan and Joni Mitchell and the Grateful Dead, then songs I make up. I sing to the old gypsy woman in my dreams, to the trees, to the sea, to the Native people who used to live here, to all the spirits that might be angry. I tell them I'm just visiting, that I'll be careful, that I'll try to do no harm.

For the next few weeks, ecstatic at being alone and terrified at how alone I am, I listen for the small voice: Sing, it says. Walk. Eat. Sleep. And I do what it tells me. My voice.

When Tim sends a note to ask if he can come visit, I use the public phone at the General Store to tell him I'm not ready. Still, when the wind blows and the huge trees creak at night, I'm afraid, and somehow I get the idea of a dog. I've never owned a dog, but she could be a companion, a guard who'll bark if anyone comes. So the next time I'm in Vancouver I pick up a Buy & Sell and return to the island with a six-month-old sort-of-beagle puppy. I give her two names: Ruby for the tough part of her, La Rose for the soft.

Ruby goes with me everywhere, and one day in October when we're hitching to the General Store for groceries, we're picked up by a guy with long blond hair and a ruddy complexion who introduces himself as Rob.

"You're new here."

It's a statement of fact. He knows because he was born and raised an islander. Would I like a tour?

I nod. He seems gentle. I'm in no hurry, and clearly neither is he. This, I'm learning, is the island way. We start toward the commercial area, but before we get there he turns right, toward the water, onto a small road, at the end of which several carpenters are building what Rob says will be the first pub on this part of the island. He shows me around, introduces me to one of the carpenters, Mike, who starts to chat before Rob cuts him short.

"I'm giving her the tour!" Rob drives us back to the main road and south across a single-lane bridge to a second island, and as he talks ("This used to be one island") he pushes behind his ear a strand of blond hair that promptly falls forward again. His fingers are rimmed with black: dirt or grease, I can't tell. After the tour, I invite him back to the cabin for tea—my first visitor.

"Living on an island is like living on a boat with a big family," he says as we settle at the picnic table. "None of us chooses the people who are here with us, but there's nowhere to go, so we all just have to get along." He looks at me from under one blond eyebrow. "How long are you planning to stay?"

"Three months."

"Lots of people come for three months," he says, "and stay three years. Maybe you'll be one of those."

It's Rob who tells me it'll soon be cold, so the next day Ruby and I hitchhike to the General Store to ask about the airtight stove, and that afternoon, Dave, the owner, delivers a huge cardboard box to my cabin door. I open it and take out what looks like a large black can on four legs with two holes in its top, three pieces of flat black tin rolled like papers in a mailing tube, crimped at the ends, a small silver dish and a silver handle. I'm going to have to assemble the thing. Panic.

There are directions—four cartoon pictures on the side of the box—and I have the screwdriver I bought last summer in Fort St. James. Following the first picture, I place the can on the gravel patch below the opening in my ceiling. So far so good. Second picture shows a ring of stovepipe rising from the can, but the only other things in the box apart from the silver dish and handle are the rolled pieces of metal. By playing with them, I manage to make three tubes that—fitted together—give me something like the picture. But the tube is too long. I look over at Ruby curled on the couch, for moral support. She opens one eye, shuts it again. I'm on my own.

I can't do this. I go sit outside, waiting vaguely for someone to save me. When I come back in, I can see right away that if I detach one length and slide the remaining two up inside the roof hole, I can re-attach the third piece as I hold it. It works! But I forgot the silver dish and handle which, I know, are for adjusting the feed of air to the fire. I dismantle the tubes and, using a rusty hammer and nails I found under the house, punch two holes in the pipe. It's like punching holes through a turkey breast to sew up the stuffing. I fit in the dish and handle, put the pipe back together and voila—a stove!

I marvel at what I've just done.

That night it rains hard. I sit in the easy chair trying to read Loren Eiseley's *The Unexpected Universe*, but I can't concentrate, and after a few minutes, I hear a noise. It's not the buoy whistle, which by now is as familiar as a mantle clock. This noise starts low but grows louder, a ripping sound, as if someone is tearing the jaw off my face. Then quiet. As I start to breathe it comes again, the sound of bone bending, but not bone, of flesh tearing, but not flesh. Finally, hands shaking, I pull my clothes back on.

Outside, it doesn't sound so terrible. In fact, it's only the huge cedars and firs, swaying and creaking above me—and one low branch that rubs against the house every time the wind gusts. I tuck the branch behind another tree, then stand for a minute, looking over the water. It feels peaceful out here. The rain has stopped, the channel is lit by a half moon that flirts behind clouds and the lights of Vancouver Island flicker like drowned stars across the black water. Still, there's such a feeling of power in the giant trees above me, they demand response. I drop my jeans, squat under the largest cedar and pee.

—

I have a stove, but now I need something to burn, which is no doubt why Dave at the General Store recommended that I also buy a cute, triangle-shaped saw with serious teeth that he called a swede saw. I'd picked the bright orange one for its cheerful colour.

"Keep the blade straight!" he'd called as I left.

Easier said than done. In one of the empty lots around me, I find the right thickness of dead tree limb balanced at waist height. After some sideways skewing of the blade, I find that if I place both hands on the saw, squat like in kung fu movies and concentrate fiercely, I can cut all the way through. The branch falls with a satisfying thud, and gathering firewood now becomes a daily chore. Ruby loves these jaunts. The salal and fern on the forest floor are so thick that often I can't see where she is until a small brown beagle suddenly bounces high above the green— eyes wide, ears flying like some canine Dumbo. On those days I call her Manfred, after my favourite childhood cartoon about Gerald McBoing Boing and Mighty Manfred, his Wonder Dog.

On one of our excursions we find a stack of boards that I figure have been abandoned, and Ruby and I carry them home. I like fixing things—nailing the found boards over a rotting step, putting up a book-shelf, building a flower box on the deck. Once I even help a neighbour who's building a house and needs someone to "hold the ends of things." It's fun, but after two hours he says I must be tired and takes me for coffee.

One day while we're out walking, Ruby and I meet a woman, also with dog, who fills us in on the local lore. This isn't really a lake, she says, as we stand at one end of what looks remarkably like a lake. It was dug by the people who developed the area and is full of leeches, she says, "so don't try swimming here." Oddly—or maybe not—it's called Magic Lake. When I tell the woman that the one thing I miss on the island is a library, she says there is a library—a mail-order library called Open Shelf, run by the provincial government.

"Stop!" I yell at Ruby as she races yet again across the road in front of an oncoming car. The woman frowns as Ruby ignores my piti-ful command and races back, inches ahead of a truck. As we part, the woman mentions that if a dog chases sheep here, farmers are allowed to shoot.

That does it. I mail a request to Open Shelf, though when *How to Train Difficult Dogs* arrives in the mail, it hardly seems to fit my sweet puppy. And our training sessions are not a great success. Eventually we work out a compromise: as long as Ruby doesn't chase sheep or dash across the road without first checking with me, she can do as she pleases.

One evening a fierce storm hits. Wind screams under the door and beats at the windows. I go to bed early and cover my head with the pillow, but trees lash at the walls and Ruby whines until I start to wonder if the cabin is safe. If wind takes out a window or—worse—a piece of wall, I'll be swept away. I throw on my clothes and go downstairs. Ruby is frantic. What can I do? Then I remember the tiny grey-haired woman and her ex-logger husband I'd met a few days ago out walking—Hélène and Ernie. They'd insisted I join them for coffee at their place, a comfortable double trailer nearby, and sent me home with a bag of veggies from Ernie's garden.

"Come back any time," Hélène had said.

I put Ruby on the leash and open the door. Wind whips it out of my hand and slams it against the wall. The next gust shoves me like a fist, pushing me back into the house. Ruby and I skitter down the stairs and through the woods to Hélène and Ernie's where I pound on the door, too scared to feel guilty about waking up old people in the middle of the night. A light flashes on and Hélène appears at the door wrapped in a flowery kimono.

"I was worried about you! Come in! Come in!"

She gives Ruby a treat and leads us to the spare bedroom, where I'm put to bed beside the largest Christmas cactus I've ever seen. I cuddle in under the thick duvet and am asleep in minutes.

The next morning when I venture home after a substantial breakfast, I still have a roof and walls, but the roofing on the east side hangs in tatters. I have no money to pay someone to fix it, but it can't stay like this. I head to the restaurant for a coffee to give myself time to think. On the rare times I've come here before, I've been careful to catch nobody's eye, but today Mike, the carpenter Rob introduced me to, sits at the next table. When he asks where I'm living, I tell him about the no-plumbing-no-water-no-telephone bits and, now, the tattered roof.

"What's the roofing made of?"

"Like tarpaper, only thicker."

"Roofing felt," he says. "I can help you fix it, if you like." And he gives me a list of what I'll need. "If Dave saw your place when he delivered the stove, he'll know quantities."

That Friday when Mike arrives, he approves the neat pile of supplies that sits beside the house, and the steel-toed boots I'm wearing, from piling lumber last summer. He roughly shoves something leather at me. A carpenter's apron.

"It's an old one," he says, not looking at me. "I don't need it anymore."

I wrap it around my waist, adjusting it to the smallest hole, then the two of us gaze up the twenty-foot slope of plywood to where the top layers of roofing have ripped off.

"Ladder?"

A quick detour to Ernie's and we have a ladder that reaches almost to the peak. For someone so well educated, I have no idea how we're going to juggle heavy rolls of roofing felt, nails, paintbrush and tar twenty feet up in the air, but I love the fact that I'm outside on a clear fall day, perched at the top of a cliff over a fabulous view, about to find out.

I do exactly what I'm told. Mike holds the ladder steady as I climb up first to tear off the loose paper. I go slowly, getting used to my boots on thin aluminum rails. When I reach the top, I catch my breath, but if I lean my whole body forward, I can feel the solidity of roof beneath me. The smell that wafts out as I tear off the old felt is almost like bread, and after a few minutes I'm too busy to look down. In fact, it's fun. The roughness of it, the big body motions, the resistance of felt and nails, remind me of piling lumber.

Mike works out an ingenious scheme using coat hangers to hold the roofing felt, and we're soon finished the first row, with me holding the ladder while he tars and nails. When we come to the second row, he says, "Your turn," and I go up, the awkwardness of the felt roll beside me, the heaviness of the tar can in my hand.

"Every four inches!" Mike calls.

But when I place the fat nails between my thumb and forefinger, the way I've seen men do, I hammer my fingers.

"These nails are too short!"

"Hold them between your two first fingers, palm up."

It works, though the back of my left hand is soon raw from rubbing on the asphalt, and I don't care that it takes me four taps where it took Mike one. I'm nailing!

I make sandwiches for lunch and we work for another couple of hours until my roof is a smooth patchwork of black and green. I thank Mike profusely, aware that I now owe him a debt, which—I promise myself—I will not be paying back socially. "Let me know if you need any help on your place!" I call as he leaves.

"Sure," he says, and smiles as he turns away. I call Ruby to come to the east side of the house with me to admire it, and the next night when there's a heavy rain, Ruby and I stay perfectly dry.

—

I have a new relationship to the elements, starting with water. It comes from having to carry every drop I use. One sunny Monday I declare a wash day and spend the morning on my deck, sloshing clothes and sheets around in a white plastic baby bath—another treasure from under the house. Slowly I strip until I've washed every piece of clothing, and I end with having my first shower, the entire tub poured over my head in one delicious warm gush. You can't appreciate hot water until you have to heat it yourself. After, I lie on the moss at the edge of the cliff, throwing sticks for Ruby and admiring how handsome my laundry looks hanging on a rope I've tied between two trees.

Maybe it's the familiarity of laundry, or the sun coming out, but today—six weeks since I arrived—is the first day I haven't felt afraid. As I eat supper at the oak table, I watch the sun set in a bowl of hills across the water. Every sunset is different, and spectacular. Tonight it's a blue one, no other colour except to the west, where fog moves slowly to embrace the island in front of me.

Still, nights are hardest, and when another storm sets the big trees outside to a frenzy, I give up pretending to read and go outside. The wind tonight is a wild thing that whips the water into whitecaps while the trees moan. I approach the big fir beside the deck and slip my fingers into the deep crevices of its bark. Holding hands with a tree, I feel sheltered and oddly warm. As branches hiss and roar above, I lean my ear and cheek against the trunk and listen to the sibilance of fir and pine needles and the lighter swish of alder. Gradually I become aware of a hollow sound, as if I'm pressed not to bark but to the steel hull of a ship. It's comfortable half-lying here, drowsy, with my full weight against the tree. For a moment I think I must look silly, but there's no one to see. I feel my body tickle as if lining up, blood running parallel, then inside the hollowness I hear a hum, a faint musical note, a single string like a cello's, and fainter, a deeper note crossing it. When I go inside, my fingers smell of pitch.

It's getting colder, which means a whole new relationship to fire, too. Fire must be wooed, and it's Rob who shows me how to light the airtight using moss and finely cut cedar as starters, and a long twist of burning paper reached into the can instead of a small match. Bark is easy to find around the cabin but it's fire candy that disappears fast, and no matter how long I work at collecting branches, there's never enough wood. So when Rob drops in for tea one day, bringing his chainsaw, I'm profoundly grateful. He cuts up a tree fallen on my property and splits the heavy rounds of wood while I carry and pile until I feel warm just looking at the neat rows beneath my front window. I make us grilled cheese sandwiches, and Rob brings out a small, potent joint. Whenever I try to sound intelligent with Rob, I come off sounding like a school-teacher, so mostly I listen. I'm fascinated by the way he talks, unlike any city man I know. Today he's excited.

"I have a great idea—I'm going to build a workshop thirty feet up in the firs on my property. With metal siding—like a space ship. Cool!"

He leaves after a pleasant afternoon, and that's the first night I forget to lock my door.

Regularly now, when I get lonely, Ruby and I go to Hélène and Ernie's for coffee, and perhaps it's Hélène's delicious pastries that inspire me to buy a tiny electric oven for fifty cents when I spot one at a garage sale on the island. The oven knows only High and Off, but I make bread anyway—hard as rock, but bread. I love the gooey texture of dough on my hands, the smell of yeast and bread baking.

Another great pleasure is mail; every day I walk to the end of the road and open door number 49 in the green metal post box. Mostly, the letters are from my family. Once there's a care package from Mum holding a new set of towels, real work socks—grey wool with a red stripe at the top—and homemade cookies. A week later, after I've returned the favour with a box I called "Kate's House," packed with curly red arbutus bark, one of Ruby's chewed sticks, a loaf of my bread and a hand-knit potholder, I phone to ask if they've received it. My father interrupts to ask, "So when are you going to start work?"

My dad can deflate me faster than anyone I know.

"I am working, Dad," I tell him. "I'm gathering wood! And in the spring I'll go back to school." Still, I must have talked a lot about the cold, because a few weeks later in the mail I get four double bedspreads: "Leftovers," a note from Dad says, from his latest show of linens. It's

Mike who suggests hanging them as curtains to keep the heat in, though I save the brown and white batik with matching sheets and pillowcase for my bed. My cabin is beginning to look splendid. But cold. It's barely December and I'm burning wood at an alarming rate. Several times, Rob or Mike has come by to help me cut and split the blocks they've left, but one day as I pass the pile in my clearing I think, if I want to live alone, I need to split my own wood, not wait for some man to do it for me. There's an axe under the house. I go and get it.

When Rob and Mike split wood, they first set a round on the chopping block, then lift the axe over their heads and bring it down in a smooth swing that splits the round neatly in two. When I lift the axe, it's so heavy I can barely get it to my shoulder. When I let it fall, I almost whack my knee.

I've worked one log into a neat pile of splinters (good for kindling, I tell myself) when I hear, "Hey, Kate!" It's Klaus. We've met briefly at the restaurant. I'm in no mood for a visitor, but he's spied the axe, and before I can say anything, he leans over and takes it from my hand. I hate it. Or at least, the independent-feminist-I-can-do-this-for-myself part of me hates it. The other part, the pitiful-ignorant-poor-me part, lets go of the handle.

"I was a logger until I wrecked my back. Want me to show you a few tricks?" At least he's gracious enough to ask.

In a single motion, Klaus swings the axe to bite into the top of one of the big logs and lifts it onto the chopping block. I can't help it: I'm impressed. He shows me how to stand square to the block, legs braced, arms stiff, axe-head square to where you want it to hit the wood. He looks like a warrior. "Don't fight the wood. Check the grain and aim for the weak points."

He swings, like one of those golfers on TV, lets gravity do the work as the axe falls. The log splits obediently into two pieces that tumble off the block. He cuts several more, then hands me the axe. "Your turn."

The first time, I get the axe firmly stuck. "At least it's in the centre!" I say. Pitiful. But he shows me how to put my foot on the round so that when I hit the axe handle sharply with the side of my fist, the blade slips free.

When I get it stuck a second time, he says, "You're aiming for the top of the wood."

Well, of course.

"Aim for the bottom," he says. "You're going all the way through, remember."

The next time, I swing high, aim for the point at which my wood meets the chopping block, and bring the axe down with all my force. I'm a banshee, an Amazon, Wild Woman—and the wood parts like water for an Olympic diver, one piece tumbling neatly to each side.

From then on, I split my own wood. Before I start, I take a second to acknowledge that this log is about to open for me, to keep me warm, and almost always, the blow splits the block first time.

One afternoon after a long walk and an hour spent splitting wood, fog rolls in so thick that Ruby and I float in a white bubble bath. Ferries pass somewhere below the cabin, blowing their heavy horns between nowhere and nowhere. We're bundled snug in our cabin, with a fire crackling beside us. Before dropping another piece of freshly split log into the airtight, I put my nose to the wood and the smell is so sweet, I'm almost dizzy. Later, lying in bed, I think about how much I like being here. I know four people on this island, which is plenty. I discourage city friends—including Tim—from visiting. I sleep long hours, eat well, meditate and do yoga each morning, smoke dope during the day, drink brandy before bed, work at gathering wood and reading, and sit naked on my front deck, singing, when the sun shines. What could be better?

One night in November, after I've spent the day with Mike and a few others cutting wood and drinking rye, he drives me home and we kiss goodbye. And it happens. Then we have to recover because it really is—we both know—too soon, and neither of us wants to wreck a lovely friendship. After we put our clothes back on, we talk a bit too feverishly about Scrabble, then kiss again. I like his smell. I like touching him. I like how he touches me back. The next day on the deck I read Herrigel's *Method of Zen* but—I hate to admit to myself—with one ear peeled for Mike. I like his cynicism, his practicality, his silly sense of romance— and his love of books.

"Help yourself," he'd said one day when I was at his place looking at his book shelves. I'd picked a play by Henrik Ibsen, *Hedda Gabler*, and that night, as usual, I sit in the armchair and read.

Hedda Gabler is an intelligent, nineteenth-century woman as confused as I am. She's locked herself—or been locked by her time—into a traditional role where the only way for a woman to accomplish anything is to manipulate a man into doing it for her.

I pull out my small stash of dope and smoke a joint.

Men are meaningless to Hedda, interchangeable.

Now I'm thirsty. I fix myself a hot toddy.

Like me, Hedda has rejected having a child. In the end, her one direct act—her own death—is staged with no other purpose that I can see, except to gain her freedom from a man's power. Her death also has the desperate overtones of abortion, though Ibsen doesn't say that outright.

The play scares me. I want security and a steady man one day, sure, but right now I want to feel my own power in the world. I don't want to live through a man, or for a man, but even in the twentieth century, what else is there? I mix a second toddy, thinking about balance. Mine, as I get up to poke the fire, is none too good.

—

One afternoon as Ruby and I are out walking, a bright yellow VW van pulls up.

"Hey, neighbour!" It's Gordon, whom I met a few weeks earlier in the restaurant. He's tall, slim, with a mustache over a mouth that seems to be always smiling. In the passenger seat beside him is a blond woman with long hair floating to her shoulders.

"Valerie and I are off to get a load of wood. Interested?"

Later, with half a truckload of fresh wood dumped in my clearing, we go to Valerie's for supper where I meet her friend Jeanne, Jeanne's partner, Stevan, and their baby daughter, Miriam, and we all (except the baby) get nicely stoned on marijuana and home brew. After that, we often get together for potluck suppers—and we eat very well. Gordon might bring a whole wheat and spinach noodle tuna casserole, me a cabbage, cheese, sesame and sunflower seed salad, Valerie her delicious dark bread, and Jeanne some of the best soups I've ever tasted.

In fact, we're all so into good food that when the only restaurant on the island closes, Gordon, Jeanne and Valerie take it over. All of us help to get it ready: painting, building benches and booths, and sewing blue-and-white-check cushion covers. On opening night, dozens of Jeanne and Stevan's homemade candles give the room a warm glow. The men who've brought their instruments play lively Celtic tunes, and Gordon makes a speech thanking everyone—especially Michael, for the beautiful wooden sign he's carved, which spells the name of the new restaurant, "Huckl'berry." The fact that it has only one "e" makes it all the more special. Everyone cheers and Michael blushes.

Lately, all the women on the island seem to be flirting with each other. For a long time four of us dance in a circle, lost to music and the pleasure of being together. One of them is Joanne. At the new pub, she and I have had long talks about being single women, eldest daughters of large families. Both of us, we've confessed, keep forgetting to shut up when we're with men, to say yes, to go along. For that alone, we love each other. And during the dancing that night, something changes for me; for the first time, I feel a part of this island. Joanne is key to that. When she drives me home, we park on the road outside my cabin and kiss and kiss and kiss. That's all. That's enough, as if we've now sealed the pact we both felt with each other, with this place.

In mid-December, I lock up the cabin, buy a travelling cage for Ruby, and leave on a six-week trip to visit my sister and her family in Calgary, then my parents and youngest sister and brother at home in Montreal. I bring Mark Vonnegut's *The Eden Express* with me. I still lean to books about madness; I take courage from the fact that others have survived.

Being home after two years away is like conducting an anthropological study. I watch my father with new eyes. For all his bravado, he's actually shy, charming, sensitive, using his drug—alcohol—to be the life of every gathering. If my father isn't the centre of attention, he fades away. Opportunities for argument are still plentiful, but when I hold back, I notice that my father, my whole family, are incessant talkers; it leaves no room for pain. But there's care here, too, fun with my brother and sister, and even—for the first time—with my mother. One day as we're watching television, I hear an unfamiliar sound.

"I've never heard you laugh before, Mum!"

She barely glances away from the TV. "Never had time."

My Christmas presents are a pair of leather work gloves and the chainsaw I've asked for. And a bottle of perfume, because "we couldn't just give you a chainsaw!" as my mother explains. I'm touched that, through their worry, my parents acknowledge, even accept, what must seem to them the rather strange lifestyle I've chosen.

4

head down, ass in the air

When I signed the lease on the cabin, I'd planned to stay for three months, but my money keeps stretching. It's cheaper to live here than in the city, and when I get back to the island early in January, it's been four months and I'm still not ready to leave. Sooner or later, though, I'll need work and there aren't a lot of jobs on an island of nine hundred people. It's not pressing but I start asking around: any jobs for a secretary? waitress? One day I catch a ride with a man who mentions needing someone to help him lay carpet.

"I'm looking for work," I say on impulse.

He looks at me for a long instant, then shifts his gaze back to the road and doesn't say another word. I drop it.

In March, on the day I turn thirty, I wake up in the night to check the clock. For years we've all agreed that you can never trust anyone over thirty. Thirty meant old-fashioned, conservative, practically dead. Now I lie in bed feeling relief, because no matter what thirty is, I'm it. The joke is that I don't feel conservative at all. I don't feel stuck. I feel like a skinny little branch about to burst into bud.

By June, I can't put off the problem of money any longer. I begin to think about leaving, maybe going north again. At a lamb roast at Joanne's to celebrate solstice, I join a circle of men I know who are talking about how much they love this island, how cheap it is to live here.

"But there's no work," I say.

"I just quit my job at the school," says a tall, blond man I recognize from the bar. Ken.

The biggest job on the island right now is building a new school and community centre. It will take all summer, and half the men on the island already work there.

"So why don't you apply for a job as a carpenter?" Ken continues.

I look at him to see if he's joking. The idea's unimaginable for a hundred reasons, but I say the first thing that comes to mind. "I don't know how to build anything."

"No problem," another man says. "Lie."

"Yeah," someone else agrees. "Tell them you worked somewhere they can't check."

The others nod, as if this is normal. I wait for one of them to point out the obvious—women don't, women can't do construction.

Someone says, "I'll lend you a hammer."

All around me a small circle of gentle men are nodding yes. What can I do but nod back?

⚊

The next morning I'm not so sure. My stomach is on fire. I lace on my steel-toed boots and loop Mike's tool belt over my shoulder. Go to the foreman's shack, the men had said. I've seen it from the road—a tiny hut barely bigger than an outhouse. Behind it, the new school looks like a huge wooden box with no lid. Inside the foreman's shack, two men are bent over big white sheets of paper.

"I'm looking for work... as a carpenter." The men glance at each other. One brushes past me and out the door as the other extends his hand.

"Ray Hill. I'm superintendent here."

"Kate." I stick my hand out too. Do I do the lying now?

From a shelf beside the desk, Ray takes a blank index card. I expected more dirt. He wears a clean shirt and trousers: no hard hat, no baseball cap. There's a quiet air about him as he hands me the card and one of the three sharp pencils that lie precisely parallel on the desk. I can see now that the white sheets are building plans.

"Write your name, address, phone number and construction experience."

For experience, I write, "Houses in Fort St. James," and when I tell him I have no phone, he says, "Then you better come back tomorrow."

"I have all the carpenters I need," he tells me the next morning, "but I could use a labourer. Interested?"

I'm quivering with adrenalin.

"You can start Monday. We work 7:30 to 4:00 but starting on Wednesday we'll work 5:00 a.m. to 1:30."

"A.m.?" I repeat stupidly. "5:00 a.m.?"

"To avoid the heat," he says in his calm way, as if this is normal. I decide on the spot I shall cut back my body motions to match Ray Hill's stillness. This is how carpenters are. But as soon as I reach the edge of the yard I start to run, elated. There's only one problem.

"What does a labourer do?" I ask a man I know at the Huckl'berry Inn.

"Easy. Just do what you're told."

But what if the foreman finds out I don't know what a joist is? Joint? Hoist? The man says I'm the first woman hired for a construction job on this island, ever. At least I'll make great money: $6.18 an hour.

Years later, Ray will tell me that when he'd gone home that first night, he'd said to his wife, Beth, "You'll never guess what happened today—a woman applied for a job as a carpenter." He'd known from the way I stumbled across the yard that I'd never been on a construction site—ever. "What shall I do?" he'd asked Beth.

"Hire her."

Before Monday, though, there was my father to contend with. He'd written that he was coming west on a business trip and wanted to visit me on the island. My father never made any bones about being an urban man. All through our growing-up years we'd never had a pet (he particularly disliked dogs), he refused to take us camping ("...and sleep on the ground?") and if anything at home needed repair his line was, "Call the man." Most of the time it was Mum who painted and wallpapered and took the plunger to plugged toilets because most of the time, we couldn't afford "the man."

Now this dad wanted to visit me. He'd take the Saturday morning ferry, he told me on the phone. Had he read my letters? Forgotten I had a dog? No running water? I'm touched he wants to come all this way to see me, and terrified that now I've finally found a place I love, he'll hate it.

The good news is that I have a truck. Mike is on an extended visit to family in California and has asked me to look after it for him. It's my first vehicle: a 1953 Chevy half-ton pickup, rust coloured with white patches on the fenders. All I have to do is to keep it going—which may be a challenge. The radiator pours water, the oil leaks, the passenger door won't open from the outside and the distributor cable is on the verge of non-being. The battery is weak, the gas gauge doesn't work, and the truck won't start except from a running position, so you have to park it on hills. Luckily, I live on a hill. The power system is overvamped (or something), so I have to drive with the heat and the lights on. But it has character—wrap-around windows in the cab and fat fenders that make it look like the auto equivalent of a Reubens painting. I call her—it—Person.

So on Saturday morning I plan to pick my father up in style. Except that I wake up late and have to hitchhike because the truck's battery has finally died. Dad assumed—reasonably, given my history of forgetting family members at various airports—that I'm not coming, so he's phoned the only number I could give him, Ernie and Hélène's, and by the time I arrive at the ferry terminal, Ernie's wandering through the crowd yelling, "Mr. Braid? Mr. Braid?" I introduce them, and we climb into Ernie's car while Ruby tries to lick my father's face.

"Ready?" I ask Dad, too brightly. After lunch, when I inform him we have to get the truck's battery fixed, Dad surprises me by offering to drive. "I haven't driven a stick shift in years."

"Um, it doesn't start very well," I tell him. "You have to jump it."

Dad does an impressive job of putting in the clutch, letting the truck run to just the right speed, slipping it into second and popping out the clutch so that suddenly—motor on—we're cruising.

"You've done this before," I say with new admiration for my city-slicker father. He's concentrating now, shifting gears. Then the truck stops. Not on a hill. So we push—or rather, Dad pushes, in his Gucci shoes and expensive suit, and I steer. When I look back and inquire anxiously how he's doing, he shouts, "Just steer!" Ruby is in the back, nose stuck over the edge to catch the breeze, ears flapping whenever we go fast enough to create one.

We find Rob at the Huckl'berry Inn—the Huck, we call it—and he thinks he can find me a six-volt battery. "In the meantime," he says, "you have no lights, so you better stay off the road tonight." So I invite Ernie and Hélène for supper and Ernie doesn't stop talking. It's the first time I've seen my dad silent for three straight hours, and he and I don't even come close to arguing.

That night I sleep on the couch so Dad can have my bed, though the batik spread and matching sheets that looked the height of luxury yesterday seem suddenly meager. For breakfast, I've bought us a treat—bacon, eggs and orange juice—and I set the table inside so he'll be more comfortable. I've never known my father to appreciate scenery; he's too busy. So I'm amazed when, as I finish cooking, he says, "Quite a view." Dad seems to have gone through some kind of transformation.

After breakfast, I heat water for him to shave. As he stands in front of the five-inch mirror, twisting his neck to see under his chin, he says through shaving foam, "Interesting place you have here."

Here it comes. I brace myself.

"I remember the first house your mother and I lived in after we were married, the place where we had you." He rinses his razor and adjusts himself to find the other side of his chin in the tiny mirror. "There was no toilet there, either."

He finds it quaint! When he leaves Sunday night, Dad and I have not said one angry word to each other for two days. A miracle.

Only now do I have time to think about my new job, starting tomorrow morning. I make my favourite lunch—tuna sandwiches on croissants—and set my tool belt and boots by the door. Ray had said to bring a hard hat. "Though most of the men," he'd said, with only the slightest quiver on men, "don't bother with the hard hat."

At six, when the alarm rings, I'm up, forcing myself to eat while my dog bounces with excitement. Twice I change my clothes, deciding on jeans and a sober-coloured T-shirt with a loose shirt on top. I will look as much like the men as possible. I tie Ruby on the deck, then hitchhike.

By the time I step onto the site at 7:15, I feel like I've had four cups of coffee. Several men lean against piles of lumber near the road and others stand scattered near the shack. It helps that I know a few of them, including Rob. And Rick, the guy I've been seeing lately, though I haven't had time to tell him I'm coming to work here at the school. Rick arrives at 7:30 just as someone rings a dinner bell, but he doesn't notice me, walks straight past and on into the building. Never mind; I'll say hello later. Within seconds everybody except one young blond guy and me have disappeared into the school. After a few minutes, Ray appears.

"Cam," he says to the blond man, "meet Kate, the other new labourer."

So they've hired two of us. For the next hour it's, "Cam, get the truck." "Cam, bring 2x10s." "Cam, we need shingles," while I do cleanup. Girl's work! When I notice no one else is wearing a hard hat, I ditch mine. After an hour or so, Ray says to me, "Go help rig up trusses."

I don't know "rig," I don't know "truss," but I do know how to obey orders so I follow his pointing finger to the playing field beside the school where a heavy-set man and a skinny one are standing beside a tall pile of lumber. I lift each boot carefully as I walk, trying to look as if I belong here, but as I draw close, both men look alarmed and the heavy one, pointing above me, yells, "Heads!"

I look up to see a giant hook plummeting toward the bull's eye of my bare head. I jerk forward, trip on a clump of dirt and barely catch myself by running, hands first, into the lumber. The pile sways precariously as both men grab at it.

"Hi."

They don't answer. The skinny one flings himself against the pile— a series of huge wooden triangles—while the other catches the rope that now dangles from the crane's hook. Already, as if by magic, one of the triangles has risen and hangs quivering above our heads. Everything is happening too fast.

"Rope!"

I see a flash of yellow and grab. The momentum almost knocks me over, but somehow I've stopped the triangle from swinging, and for a second, feel absurdly pleased. Then the crane lifts and the yellow rope burns through my hand until I let it go. The heavy one yells again.

"Get out of there while he lifts!"

As I stumble away, Slim lifts his tank top to wipe his forehead and I catch a glimpse of hairless midriff, ribs finely outlined in muscle.

"Ray sent me to help you," I say, keenly aware of the irony.

The heavy one is Bruce. The skinny one is Dale. They explain that the triangles are pre-fabricated rafters ("trusses," they say). Our job is to tie ("rig," they say) each truss to the crane, which will then lift them to the top of the walls where the carpenters wait to catch them. I look up to see stick figures balanced on top of the school and the thin pointer of an arm falling and rising again before I hear the sting of a hammer.

The next time the yellow rope comes down for a truss, Bruce tells me to tie a bowline.

"Bowline?"

He shows me, but his hands move too fast. When it comes my turn, my bowline doesn't look at all like his, so I use a reef knot, the only knot I remember from Girl Guides. No one seems to notice, and the truss rises and swings to the roof as if this is normal, as if women tie reef knots to trusses every day.

In between the time the crane swings a truss up and returns for another, we wait. Which makes me nervous. You'd never be caught just standing, doing nothing, at any "woman's" job—typing, reception, even child care—yet these men seem unconcerned, punching each other's shoulders, banging their hammers needlessly against the lumber. I feel like a teenager, expected to giggle while the boys horse around.

During one of these waiting times Ray suddenly yells over that if we have nothing better to do, why aren't we picking up garbage? Frantically, I scavenge. For a few minutes the two guys do the same; then they retreat to the shade of a nearby tree while I keep going until I'm picking up twigs and pine cones. When Bruce calls, "No point in knocking yourself out," I join them in the shade—keeping an eye out for Ray.

When the coffee bell rings, I follow Bruce and Dale to where we've left our lunches on a pile of plywood by the road. Still no sign of Rick. I feel like the new girl at the party, not sure where to sit, until Rob settles my dilemma by moving his lunch box beside Bruce's and I sit down beside them. Hardly anyone talks, and going back to work is a relief.

As the morning progresses, I seem to have an uncanny ability to be in the wrong place at the right time. I'm in the exact spot where the hook is descending, precisely where a truss swings wild. My feet feel huge. Just before lunch, trying to look busy, I think I'll neaten things up by straightening the pile of trusses. When a gentle pull has no effect, I brace my feet and heave until the whole pile suddenly shifts and I fall back, twisting my ankle. For the rest of the day, every time I take a step there's a bolt of pain up my leg, but I'm not about to let anyone see.

At lunch, when I sit down with the other labourers, I have an overwhelming desire to lie back and rest. I don't, because I'm afraid it isn't what real construction workers do, afraid it will show my breasts. Instead, I try to follow the conversation. There's some talk of sports, then Rob says, "Last night, eh?" and the others laugh.

"You bet!" More laughter.

"Twenty-four, eh?"

Rob, this man with whom I've had hours of intelligent conversation, is suddenly speaking in sound dollops, and everybody except me finds it funny. The wisp of pleasure I'd felt at fitting in evaporates.

After lunch Ray sends Bruce and Dale and me to build trusses. I do what they do, carrying armloads of cut-up lumber from the sawyer's table, trotting single file like ants down one side of a ditch, up the other and across the field to dump the lumber beside a piece of plywood laid out on sawhorses. When a skinny man in white coveralls notices us staggering across the field, dropping bits of two-by-four, he snaps, "Find yourselves a system!"

"That's Jim," Bruce says under his breath. "Foreman."

"One of you lazy sods, go get a wheelbarrow!"

When Jim lays a plank across the ditch, one of the carpenters shakes his head. "That guy will find an easy way into heaven," he says, as—from then on—we labourers file in a straight, efficient line across the ditch, one of us bringing lumber in the wheelbarrow while the rest lay wood into a pattern laid out on the plywood. "Jig," Jim calls it. The jig is like a fill-in-the-blanks questionnaire, only we fill in the blanks with wood. Then we lay metal plates with holes in them at each corner of the jig and someone says, "Now, nail like hell," and I do, one nail through each hole, using both my hands on the hammer. Each time my tool belt empties, I fill it from a cardboard box of nails nearby. What could be easier? But I'm worried about the size of the nails. These stubby things couldn't hold together a doll's house. When I say it out loud, there's a pause, then Bruce assures me the nails are "engineered."

All afternoon I nail. There are three outside corners and three inside corners, twenty-four nails per corner, 144 nails per truss. As soon as one is finished, it's lifted onto a pile beside us and more wood goes on the jig, more nails. The men take one or two swings per nail while I take seven or eight, using both hands. Even at that, I often bend my nails. And it's starting to hurt. When I look at my right palm, the carpenter who's working with us, André, notices I'm getting blisters. He lends me a glove, then says quietly, "Eef you nail from your shoulder eenstead of your wreest, you won't have to heet thee nail so many times. Would you like a leetle break?" But no one else is taking a break so I say no, and do as he says, swinging the hammer from my shoulder. It feels awkward, but after a while I'm taking only four swings per nail.

I like sitting here in the sun with a bunch of men, building. I like the feeling of production as I sit, straddling the pile of sun-warmed lumber between my legs, reaching with my left hand into my leather apron (funny they call it "apron") for nails, each one identical to the last. All I have to do is put each nail into its hole and drive it home. That's what one of the other labourers is yelling, "Drive it home, Bruce, baby!" I don't even have to smile. All I have to do to make the boss happy is to nail faster.

A half-hour before quitting time, my hammer takes on a life of its own. First it gains weight, so it takes a huge effort to lift. Then it develops an alarming ability to twist in my hand. I end up flopping the side of the hammer, rather than the face, onto the nails, which bends them. Once I watch with horror as my hammer sails out of my hand, barely missing someone's knee. When the bell finally rings for the end of the day, I gather my pack and hard hat just as Jim, the foreman, walks by. André grabs my hand and turns it over.

"Look, Jeem." My palm oozes with blisters.

"That'll teach her to do a man's work," Jim says in a voice as pinched as barbed wire.

After a moment André replies, "In a week she'll be tougher than you!" André's trying to make me feel better, but Jim makes me mad. I'll show him. Still, by the time I reach home, I can barely crawl up the steps. My fingers are like raw sausages as I struggle to release the knots that hold the squirming Ruby. I let her lick some of the salt and grime from my face, wincing when she rubs against my palms, then limp into the house to unbutton and unhook, letting boots, shirts, jeans lie where they fall. I crawl up the ladder and collapse into bed before Ruby can whine an objection. When I wake, an hour later, I walk slowly to the lake for a bath—leeches be damned—then home again to throw together the fastest dinner I can think of, rice and stir-fry with yogurt. At eight o'clock, just as I'm heading for bed, Valerie and Gordon and Jeanne arrive with brandy to toast my first day. Jeanne takes one look at my hands and goes for the Dettol bottle, ordering me to soak them every night. By the time my friends leave, my dishes are washed and tomorrow's lunch is made. After a little yoga, I crawl up the ladder and fall instantly asleep.

At work the next day, Ray's wife, Beth, comes to take a picture. As the crew all squint at the camera, Jim calls, "Should we be topless?" Hardly anyone laughs. In fact, everyone is kind. Only Rick ignores me. In fact, Rick never speaks to me again, but I'm having too much fun to care.

On Wednesday when my alarm rings at 3:30 a.m., I'm surprised how easy it is to get up. This week there's been a filling moon and for a few minutes I watch it hanging golden-orange over the water. When I leave the house an hour later, it's barely light.

Ray starts me out sweeping. "Everyone takes their turn sweeping," he assures me, and sure enough, after the broom work he gives Cam and me each a pair of gloves and waves for us to follow him to where dozens of brown-paper-wrapped shingles lie in piles beside the school.

"We have to get these up there," Ray says, pointing to the roof. "The crane can lift them but I need you to load them onto pallets"— he points at some wooden platforms—"and unload them up top." The paper wrapper on the three-foot long bundles of shingles says, "Seventy pounds." He turns to me. "Can you keep up?" Which pisses me off.

Cam and I load the first pallet together, picking up each bundle and swinging it onto the pallet by its plastic strap. It hurts my back, but the gloves protect my blisters—mostly. I push myself to keep up to Cam. When we finish loading the first pallet, he goes up top while I start the second one by myself. My back is killing me. When I glance up, I can see that Cam, who's thinner and shorter than me, already has the first pallet unloaded.

Ray's right. "Girls" aren't strong enough to do this. I can't keep up.

The next time I slide another bundle off the pile I'm too tired to lift with my arms, so I just hug it close to my body as I swing around and drop it onto the pallet. My back doesn't hurt. I repeat the trick. It isn't the way Cam does it, but it works. I find a rhythm: pull the bundle toward me, hug, lift, swing, drop. If I concentrate on my hips, settling the weight there, then the bundles seem even lighter. It's like carrying a kid.

Three pallets. Four. My arms, my shoulders, my legs—everything burns with effort, but every time I start to slow down I think about Ray and hoist another bundle. Finally I'm done. Each of my arms has a dead weight attached, but I've kept up! I wait for Ray to admire what I've done.

"Now take the insulation inside," is all he says.

Our last job for the day is to carry eight huge wooden beams into the gym. When I see them, I'm scared—they're the size of telephone poles.

Bruce says, "These babies must weigh two hundred pounds!"

On the first trip, James—one of the other labourers—directs us. Bruce and Dale razz him for playing foreman but we all do as we're told, placing a short length of two-by-four under each end of the beam. James and Dale stand on each side at the front, Bruce and I at the back.

"All together, now," James says. I bend with the others, brace, and lift. As I straighten, both hands gripping the two-by-four, I feel more weight than I've ever felt in my life. But the weight isn't impossible; in fact, it's lighter than I expected. I concentrate, careful not to trip. By the third beam, we're moving smoothly together. As we lift the last one, I notice a young woman walking along the road beside us. Dale whispers loudly, "Should we whistle?"

I miss a step. Always before, that's been me on the road. Now I'm a construction worker, what should I do? No one whistles, but I worry about it all the way home. Part of me feels guilty that with me there, the men can't do what they've always done. Then I get mad. Why are they whistling anyway? Cat-calling women is a rude, stupid thing to do.

—

Cam and I are busy passing plywood sheets though a window to Stan and James inside when I notice my hands. In just over a week of work, my fingers have become thicker and all my fingernails have broken off. Between small burns from the airtight stove, nicks from the swede saw and getting my fingers pinched and chipped here at work, my hands are not a pretty sight. While we wait for the two inside to have a smoke, I hold up my hands in front of me like two lumps of meat.

Cam notices. "You're the strongest girl I ever worked with."

It's meant as a compliment, but it makes me uncomfortable. It's worse when, a few minutes later, the plumber stops to watch and says to me, "Don't work so hard. If you work too hard, you'll get muscles. Nobody likes a girl with muscles."

"Don't pay any attention," Cam whispers when the man leaves. "He's just a sexist."

Cam is easy to work with; like me, he doesn't quite fit here. At lunch, when the men pull out thick sandwiches of white bread and peanut butter, he eats only apples. "Local food," he explained to me once. One day he finds out I play guitar, and a few times we arrange to meet before work, inside the cool dark and high echoes of the unfinished gym, to play our guitars together. When I stop going, I tell him it's because 4:30 is too early, but really, it's because Cam isn't "one of the boys," and I'm afraid that if I hang out with him, I might never be, either. I desperately want to be one of them.

Most days after work I go directly to the Huck for ice cream, where everyone wants to know what it's like to work at the school. Today I tell them how, when someone started to say "Fuck," he suddenly looked at me and stopped. As if I cared.

Randy, one of the younger carpenters who's listening in, says, "These guys never swear in front of their wives, and you're a woman. The language has cleared up since you've been there."

Which seems odd, because I swear a lot myself. But I decide if the men aren't going to swear, I won't either—until one day a bunch of us, including Randy, are nailing backing onto trusses. I don't know why we're doing this but I'm proud I've been asked to help, and I do what I'm told, though I feel graceless in the confined space, hanging by one arm trying to bang big nails into tiny two-by-fours where every time I pull the hammer back to take a swing, it hits something. And it just pops out. "Shit!"

Everyone freezes.

"Shhh!" Randy says from a nearby truss. "There are men present!" And everyone laughs and goes back to work.

Jim, the foreman, has been fretting that the good weather can't last, so Ray agrees we'll speed things up; while the roof trusses are being covered with plywood sheathing on the west side of the roof, the carpenters will start laying roofing on the mostly-sheathed east side. The carpenters grumble that this isn't the proper way to do things, but I notice they grumble at everything, then do it.

The other labourers have been on the roof for most of the week, laying plywood, and now Bruce and I are sent up to help. I've handled plywood once, but never carried it. And except for the day I worked on my cabin with Mike, I've never been on a roof.

As Jim shows us how to lay sheathing, I keep my eyes down, trying to forget I'm two stories above the ground. Jim walks to the lower edge of the roof as if strolling on a golf course, picks up a sheet of plywood and brings it back to where the ply ends about halfway up the roof. He lets it fall—perfectly in place—with a sharp slap that echoes in the gym below.

"Neat!" I say.

Jim glares. "Leave a quarter-inch gap along top and bottom edges!" he snaps. "Then tack the corners!" Jim only talks in exclamations. "Nail the rest later!"

I nail ("tack") while Bruce carries plywood. It isn't as easy as Jim made it look but we manage, and the more we lay, the faster we get. Then Jim says, "We need the rest of those ply sheets on the other side of the roof, pronto!"

Now I'll have to help Bruce carry. The lower half of the roof on this side is sheathed in an increasingly narrow corridor. Where we'll have to cross at the peak, there's a wide gap through which I can see the gym floor, twenty-six feet below. A person could fall through that hole. Bruce's face is a careful blank as he picks up a sheet from the pile, walks to the peak, and pauses a second before taking a long step across space to the far side.

My turn. The sheet of plywood is huge, and when I try to stand, the wood pulls me backward. Twice I have to readjust my hands. When it's finally right—sort of—I move slowly to the peak, then stop and glance around, waiting for someone to call me to another job. No such luck. I tighten my grip and open my eyes only when my foot hits solid wood on the other side. Trying to act nonchalant, I carry the sheet to where Bruce has left his, propped over two bundles of shingles. The second and third sheets are easier. And then the wind comes up. It's a breeze, something I'd pay no attention to on the ground, but carrying a sheet of plywood even in a breeze, I discover, is like trying to carry a wall while someone scrambles up the other side. There's a puff just as I approach the peak and my plywood is a sail struggling to carry me away. I stop dead, desperately holding on. I might die here. But Bruce isn't complaining, so neither will I. When the wind drops, I take a giant step over the gap onto solid ply.

—

I've arranged with the university to finish my coursework from the island, attending only some of the classes, and one Saturday when I come back from Vancouver, Mike—who's back from California but is now living in Victoria—meets me at the ferry. He stays overnight and when I wake up the next morning, he's gone for eggs. On the table, written on one of my yellow work pads, is a poem that starts:

> went to church today
>   in your body
> sermon was on my lips

and ends, "In holding, holy are we."
I love it that he thinks in metaphor—in comparisons—like me.

—

On Monday, Ray sends me straight up to the roof. Jim's arm makes a white staccato motion against the black of shingles as he gestures for me to come over. His mouth twists in a tight grimace.
"Ever done shingles?"
I shake my head, cautious. "Not this kind."
"Well, you'll never learn if you don't try," he growls.
He shows me how to sit sidesaddle on one hip, how to lay the long

black rectangles of roofing tile, keeping them in a straight line by measuring up to the blue chalk line he's marked on the tarpaper higher up, how to hold the short roofing nails, palm up, as Mike showed me on my roof.

"You'll probably hit yourself anyway," he says smugly. When I do, and grunt to cover the pain, he shakes his head, keeps working beside me. I watch him from the corner of my eye: how he sits, how he reaches for his nails. When he notices the backs of my fingers are raw from rubbing across the asphalt, he brings a roll of black electrical tape out of his pocket and shows me how to wrap my fingers.

I soon find it more comfortable to do the work kneeling with my knees wide apart. This spread-legged position is faster than sitting on one hip, and nobody seems to notice. It's almost fun, getting into a rhythm, trying to keep up. Until Jim pauses to watch my sideways shuffle. I cringe, but all he says is, "You should get yourself some kneepads," and shows me the careful stitch marks where two circles of sheepskin have been sewn inside his coveralls. I only grunt—I'm learning to talk less—but I marvel at the ingenuity of it.

"Head down, ass in the air!" Jim calls to the crew.

He regularly makes jokes about my shirt coming off, but no matter how hot it gets, and no matter how fast everyone else's shirts come off, he always wears the same white coverall. I've heard the guys joke about what he wears—or doesn't—beneath it, though never within his earshot.

After two days on black tile in ninety-degree weather, I can't stand it. As I pack my lunch, I also stuff in a sun top, the most modest one I own. I pick my time. When everyone's busy, I climb down the ladder and find a private corner where I can put on the top, then take a deep breath and climb back up, grab another tile and start hammering. I feel naked. And I have the strangest feeling everything's suddenly gone quiet. After a few seconds Jim says loudly, "Okay, men, talk it up, talk it up!" And the hammers and small talk start again.

It's so dark when I come into the school at quitting time that afternoon that at first I'm blinded, so I can't see, only hear Jim bragging to Ray.

"If the others keep sitting on their keisters, the two of us will have that roof finished!" Then he spots me. "Ha! Here she is, the bionic woman!" And before I can protest, he pats my behind. I frown after him as he walks past, laughing. Later, I find out the guys have also named our automatic nailing gun the Bionic Woman.

By the end of the third week, my fingers ache so much that I can't string my guitar. My knees ache when I walk, my shoulders are twisted cables, and I'm serenely happy. I love being able to see what I've done at the end of a day. It's not like housework or a typed page, where whatever I do instantly disappears under dirty feet or into a filing cabinet. I love it that for the rest of my life I'll be able to drive by and say, "I built that." I love being outdoors all day, feeling my body fit and tanned. I'm making great money, and even though I eat mountains of food, I seem to be losing weight. And all my senses are alive. The smell of fresh sweat is a turn-on, and the sweet smells of fir and hemlock are everywhere. Sometimes, if I accidentally rub against a bleeding board, I'll carry the perfume of pitch with me all day. There's the scratch of boards, the prickle of insulation, the sharpness of nails. When I read Allan Watts quoting the Zen poet P'ang-yun,

> How wondrously supernatural,
> And how miraculous this!
> I draw water, and I carry fuel.

I know exactly the feeling. I knew my body was good for sex but now I know it's good for more—for doing physical work, for building things.

Still, mostly, I feel stupid. Working here is like being in a foreign country with a whole new language. A carpenter asks me to bring him shiplap, and by the time I find Ray, I can't remember what I've come for, only that it has something to do with water.

"Do we have any... um... surfboard?"

Ray is silent for a long moment. "You mean shiplap?" He points at a nearby pile. "It doesn't matter so much what you do, Kate. Just get the name right."

Picky, picky, I think that night as I chop onions and dice garlic for spaghetti sauce. But I'm using a particular language for cooking, too. I learned it from my mother. How confusing if she'd told me to broil the soup when she meant simmer! I start to pay more attention to the names—like the tools. I've never seen, let alone, used, most of the tools these men handle so casually, so, when a carpenter named Ted, working on the roof, calls for a crescent wrench, I repeat, "Crescent wrench?" He draws a picture on a scrap of lumber and throws it down. It looks like a lollipop with a mouth.

"Silver," he says. I hold the picture up to the tools in his tool box until I find it.

I get assigned to a lot of cleanup duty, but siding saves me. "Board and batten," Ray calls it—long vertical lengths of cedar that will cover the exterior of the school.

From a distance, a scaffold looks like a tinker-toy. Up close, it's made of metal tubes that fit together into squares, with more tubes set like a ladder at the ends for climbing. The planks you walk on are laid over the crosspieces. I fill my belt with nails, put both hands on the bottom bar of the scaffold and hold tight. Lift one leg to the first step. Hold tighter. The bars are twelve inches apart so it's a huge step, then a breathtaking lift to the next level. I don't look down until I reach the top. Now I'm balanced on a tiny tower with two boards for a floor. I hang on to the end bars. The air up here is thinner.

"Get over here," a carpenter orders from the far side of the boards.

I use the wall, placing my palms flat against the smooth tarpaper like a blind woman. Someone shoves a length of siding up from below, and the surprise nearly knocks me off.

"Nail it! Every sixteen inches!" the carpenter says. I let him grab the siding, leave my left hand on the wall, reach carefully for the hammer at my right side, pull it out of its loop, rest it against the wall, lift my left hand off the wall and reach for a nail as he waits.

"Today!" he snaps.

Soon I learn the rhythms of climbing a scaffold, the wide swing from side to side so it's half momentum, not just muscle, that carries me up and down. One day, just when I'm starting to feel comfortable, I take a thoughtless step backward and fall. Then stop, safe on the over-lapped plank an inch and a half below the one I was standing on. The carpenter I'm working with sees the shocked look on my face and roars with laughter.

"That's what we call the inch-and-a-half heart attack," he says.

Last week as I tripped and stumbled, I realized that one of the things that labels me "rookie" is my lack of grace. It's clear from the way I awkwardly pile lumber on my shoulder, or dig around in my nail pouch, or trip on garbage. A dancer told me once that dancers pretend they're puppets and someone is holding them by an imaginary string attached to the top of their heads. I start to practise walking as if a string holds me up, and sometimes, on the good days, I feel almost graceful.

One morning as I brush my hair, I notice something wrong: the top part of my right arm looks distorted. When I touch it, it resists—a lump. In shock, I hold up my other arm: the same lump, slightly smaller. I stand with both arms flexed like somebody lost from a body-building contest. Now I understand why my blouses have begun to feel tight. I've grown biceps and I'm not sure I like it.

5

the best man for the job

There's talk of needing fewer labourers on the job, but I really want to stay. This morning, as I'm sweeping the gym—again—I say to Ray, "I shouldn't be doing this."

"Why not?"

"Because I know how to sweep. People who come to see the school will get the wrong impression."

"Would you rather insulate?" He has an odd light in his eye.

"Yes," I say. Foolish woman.

When I crawl though the hole in the gym ceiling into the attic, I'm in a space that's church-like in the sharpness of its angles, only steeper, and much, much hotter. A small mountain of insulation is piled in the centre, and the other labourers are taking what look like little pink quilts and pushing them into the spaces between rafters. Bruce calls the quilts "batts" and shows me how, when he presses one into a rafter space, it stays there.

"Nice!" I say.

"Pressure fit," he replies.

The batts are as prickly as cotton candy. I tug at corners, fill every tiny gap, neatly. I'm wearing coveralls with a regular T-shirt and a sleeveless one underneath, but after an hour it's so hot that I have to take the top T-shirt off and the insulation quickly begins to feel less like cotton candy and more like fire ants running up and down my arms, neck and face. And I can't stop coughing.

Ray appears through the access hole waving something white that dangles from elastic bands. "You guys better wear masks," he says. "Keep some of that fibreglass out of your lungs." I cough. "So you don't do that."

But it's awkward, breathing with a paper bag over your nose and mouth—and hotter. As the day wears on, I feel more and more like a space creature, listening to my own heavy breathing, drinking my own sweat as it rolls off my upper lip.

"Having fun?" Ray asks on one of his trips to check on our progress.

"Yeah." And I actually am having fun, watching the walls slowly turn pink.

"When you're finished up here," he calls as he climbs back down the ladder, "there's more downstairs."

After work I go straight to the lake. But even with a second bath when I get home, my arms and legs and face still prickle.

—

The next time Dad comes to Vancouver, he invites me to meet him in the city. "I'll buy you supper," he says, but first we go to a reception for his clients. When Dad introduces me to a man dressed in a fine blue suit, he says, "She's a carpenter!"

"Not a carpenter, Dad," I tell him. "I'm just a labourer."

"I can't tell people that," he whispers, and greets the next customer.

—

At work these days, Ray puts Bruce and me together a lot as partners. When we're with the other labourers Bruce tells the crudest jokes, but when we're alone, he's different. He has a quiet to him that I like. I also like that he always seems to know what's coming, so that whatever we're doing goes easier, like when Ray sends James and Bruce and me into the crawl space under the school to put in the last of the insulation. Bruce shows me how to rig up a trouble light like his so I can see what I'm doing, then we all lie down on our backs and begin stuffing insulation up over our heads. Tiny glass fibres drift into our faces as we scuttle like upside-down spiders, turning over awkwardly only when we have to pull out our knives to cut an end piece. I blink hard; fibreglass in contact lenses is like bricks in your eyes. Every time I decide to get rid of the face mask so I can breathe better, I see the growing pink circle of glass where I've breathed in and put it back on.

The trouble lights attach each of us like an oxygen hose to an electrical outlet outside. We're three fireflies crawling in semidarkness with a pink glow around us, until James bangs his light hard against a joist and Bruce jerks his cord too tight just as I knock my light on a pillar. We're in total darkness. Alarmed, I sit up abruptly and hit my head on a joist with a *thunk*. I pull off my face mask, lie back and rub my forehead. The three of us lie on our backs in the dark, not talking. It feels wicked to lie still, doing nothing. And strangely intimate. We listen to the creaks of the building settling, the boom of boots on the floor above us.

"Want a toke?" It's Bruce.

Once in a while when the carpenters aren't around, one of the labourers brings out a joint with some exotic name like Tijuana Gold or Acapulco Purple. Always before, I've said no, mostly because I worry about hurting myself. But today there are no power tools, only our knives.

"Okay."

Bruce lights a match so we can find each other, then lights the joint. As I take the tiny cigarette from his hand and inhale, I wonder briefly what happens when marijuana hits glass inside a living lung.

That Friday, they invite me for lunch. Every second Friday, the other labourers have been going off-site to have lunch at the house of a guy named Bozo, who used to be a tugboat cook. At precisely noon, four of us leap into someone's Jeep and tear off. We have to get there, eat and be back to work in half an hour.

Bozo is a tall, heavy-set man with long red hair and a bushy red beard. On the table in his kitchen, waiting for us, is a huge pile of hot-dogs and a bowl of baked beans. Even before we begin to eat, he lights a tiny joint. The smaller the joint, I have learned, the stronger. Feeling reckless, wanting desperately to be like them, I take one toke, and another. Then I lose track.

Bozo watches me with not so much a leer as the look of someone watching an animal in the zoo. I ignore him and, while the others talk in noisy bursts, concentrate on lunch. I've never tasted such a hot dog—the warm, air-filled softness of the bun and the firmness of the meat—but mostly it's the mustard. Three times I take more of the bright yellow stuff that fills my mouth and spreads until the top of my head prickles, as if someone has laid peppermint on my scalp.

When we get back to work, twenty minutes late and Soooo Stoooned, Ray is waiting. I don't seem to care that he's frowning, ordering us to dig a hole. Something about drain tile. I'm too busy trying desperately not to giggle. Ray had said we'd find the drain tile one foot down, but when there's no tile even by three feet, he says we'll be here all afternoon and we settle down to dig. The hole is just big enough for one person so we work out a system: one digs while the rest watch.

Bruce goes first. He pulls off his shirt and soon we're all watching the gleam of his torso as he works, his back shining as if oiled, rainbows of sweat splashing off as he turns, bends, lifts, in slow motion. I'm as absorbed in his body as I ever was in lovemaking, but this is different. This is sensual, not erotic, the body fully engaged, in balance, doing what bodies are meant to do: move, strain, work.

Then it's my turn.

Within seconds of entering the hole, I forget about the men above me. Four feet down, and by now we've passed through dirt and hit fist-sized rock. The clang of metal on stone hurts my ears and the rocks roll back as fast as I can shovel them over the edge, so I heave the shovel out of the hole and grab stones in my two fists, firing them out of the hole by hand. Much easier. Much quieter.

When I pause to take a breath, I hear the men above me deep in a discussion of *Zen and the Art of Motorcycle Maintenance*, at ease inside this male culture, happiest in action, with some object—a cigarette, a beer glass, a hockey stick, a shovel—in their hands. I love them for it, regardless of their strange male ways. I keep my head down and heave rocks, listening to them kid each other as if I'm not even here. And for this precious moment, I am one of them.

—

At work there may be moments of fitting in, but ever since I started at the school, people have treated me differently. Strangers, especially women, smile and chat as if they know me. Now that the roof's on, a steady stream of islanders are stopping by to take a look. One woman in her sixties reaches over and gently pushes a strand of hair back from my eyes as she asks me what it's like to work here.

"Don't try and compete with them, dear," she says. "Don't let them push you too hard." I'm surprised when this makes me want to cry.

The response from men is more ambiguous. When I walk barefoot at the Fall Fair, a man I don't know says, "Now that you're a labourer, your feet must be tough as nails." Another day, a driver watching me unload his truck says, "You must have shoulders like bullets." Comments like this make me feel oddly alone. And it's not just on the island. When I go to Vancouver for a women's conference, a lesbian I know asks if I've become one, too. Before I can answer, she says, "The women in the community are wondering, because you carry yourself differently."

And it's true. I know now who I am physically, know my physical boundaries, trust that my arms and legs can hold me, though I haven't changed my sexual preference. I wonder what this says about heterosexual women, how we carry ourselves, how we feel inside these bodies.

One day in town, I spot a comic book I've never seen before, about a character called Wonder Woman. Wonder Woman may be some man's fantasy, but she becomes my hero—hera?—and I start collecting. Sometimes I don't even read the stories; I just need her picture, her Amazon determination and confidence, waiting for me on the table after work.

—

It's early September and the weather is glorious. Mike has given me the truck (whose name is now Wonder Person), and Rob and I work out a trade: every once in a while he spends an afternoon keeping it running, while I cook supper for both of us. My routine these days is simple—five days a week I work, and Saturday is all screeching brakes as I rush to pay bills, write letters, mail off my resumé for a possible job in town, clean my house and, at around three, hit a dead stop. Usually, several of us end up sitting around someone's kitchen on Saturday afternoons, smoking dope and talking. On Sunday, my day of rest, I bake—poppy seed cake or carob chip cookies for my lunches, pie for a potluck supper that night. I knit or sort photos or sew missing buttons on my work shirt or bathe Ruby, whose greatest pleasure is to roll in dead fish. After all the summer heat, we've had a few days of heavy, sweet-smelling rain that has filled the wells and eased the fire hazard. I eat lettuce and chard and tomatoes from my vegetable box and the hummingbirds still buzz around the geraniums, but the smell as I walk through the woods is subtly different, and the colours are deeper. Occasionally, too, there's a wind, another sign of fall that reminds me I have a decision to make: should I stay here, or move to the city to write my thesis? The trouble is, there's really nothing I care enough about to research and write.

At the school, everyone's saying there are going to be layoffs, starting with the labourers, which makes each day now more precious. One afternoon Ray sets all the labourers to moving everything out of the gym. Back and forth: plywood, fir, hemlock, two-by-fours, two-by-tens, electrical supplies, more hemlock—and for a few hours I have none of the feeling I've had lately of being off balance, as if the ground under me has grown thin.

All summer we've worked nine-hour days so we could have Friday afternoons off, but a rumour starts that we're going back to eight hours a day, five days a week. When I ask Ray why, he says, "Workers' Committee decision."

I've never heard of a Workers' Committee.

"Jim."

So I ask Jim, who says it's not for sure, the crew are just talking about it. So I ask around, and everyone I talk to agrees: we don't want to change hours. That Friday I take the day off to register for fall classes, and when I turn up for work at 7:00 a.m. Monday, I have to wait for an hour because Ray's already made the change: we're now working 8:00 to 4:30. Too bad, but I guess everyone decided to go with the new hours.

On Friday at noon, when we're outside eating lunch, Jim strolls by with his lunch box under his arm.

"Where are you going?" I ask.

"Home."

I turn to Ray. "How come Jim can go home and we can't?"

"He's due overtime."

I'm not the only one who's pissed off, but there's nothing we can do, the men tell me. "Jim's just like that." But it isn't fair; every one of us would do overtime if it meant we could go home early on Fridays. I make the rounds of the crew and keep track. When ten out of twelve say they want Friday afternoons off, I head back to Ray, who says, "Talk to Jim."

By now, the crew has stopped talking to Jim, so he takes his breaks in solitary. My friends at the restaurant say we should strike.

When I approach Jim, I don't call it "changing back to the old hours." I call it "working overtime." Jim, who's usually a mild-mannered guy, bangs his hat on the table.

"I don't care when you all work your hours, but I'll be damned if I'm going to get here every day at ten to seven and have some slacker roll in at ten minutes past!"

He's talking about one of the carpenters who's late every day. But we all know there's only one.

"So if we're all on time, you're okay with overtime and everyone getting off early on Fridays?"

"If you're all on time!" Jim says, shaking his finger at me.

So that's what we do, and the guys start calling me Jane Hoffa.

On the last Thursday in September, Ray lays off the other labourers and tells me there's only one more week of work. I'm shocked that I'm the last one, and a bit embarassed, because I was the last one hired. But when I ask the others, they tell me they want to go—they have plans— and deep down, I'm glad. I want to stay as long as I can, not just for the money. I love it here.

Saturday, there's a dance on Salt Spring, one of the bigger islands, and several of us decide to go. I wear a loose yellow top and my favourite skirt, made from a bunch of Dad's old silk ties that, when the ties are opened and ironed flat, make the skirt flare—perfect for dancing. Most of the women are dressed like me in loose blouses and long skirts, with no makeup and their hair left free. The men wear logging jackets and homemade shirts and blue jeans, and the odd bright piece of silk flashes, but most of the colours are richly faded, like autumn.

The dance is held in a beautiful old hall lined with cedar, with a brightly polished wood floor and several adjoining rooms where people are already setting up sleeping bags and blankets. A woman at a table near the door sells apple juice, homemade tarts and doughnuts for twenty-five cents each, as more people fill up the hall, passing joints and beer. While the musicians warm up, a woman plays tag on the empty dance floor with three kids.

The first dancers are women. We might be taking an easy walk as we bob and turn, smiling at each other, flirting a little. When the men join us the pace quickens, and we slip together into a velvet pool of marijuana and music. Some dance alone, some in groups of two or three, and some in large circles that pull to the centre, curl shut, then gently push the dancers out again, like a huge jellyfish breathing in dark water.

I haven't danced for a long time. At first I'm a clumsy Pinocchio, my joints squared as I lumber about, fumbling for a thread in the music. Then I remember that one of the labourers at work gave me a "special," for the dance. I light it, take a toke and pass it to the person nearest me. Perhaps that makes the difference, perhaps I'm simply ready, but this time when I dance, I close my eyes and there is only the music. I can almost hear the parts of my body, rusty and clanking, and I explore the jagged edges. I am a Tin Woman, but slowly, as if Dorothy has applied the magic oil, my parts come together, arms and legs, right and left, male and female, all sewn back together down the seam of my spine. Occasionally I look up and I'm dancing with someone. We smile, touch, nod hello, dance on, a mass of shimmering selves called "tribal" for rich reason. I bend my knees, my legs part, and a sudden bolt of lightning shoots down my spine and through my cunt, grounding itself in the floor below my dancing self.

We are dancing, my body and I, and nothing can knock us over.

When the musicians take a break I walk down the road, partly to find a place to pee but mostly so I can be silent for a few more minutes. But one of the men from Pender has seen me, and follows.

"I understand," he says, taking hold of my two arms and looking into my eyes. His eyes keep sliding. "Or I don't understand, but I really want to." His breath reeks of alcohol. Whatever it is that he has to say, it's urgent. I wait.

"You're not M-F," he says, and it takes me a minute. Male-Female. "You're not M-F," he repeats, "but a person."

My ease over the past few hours teeters.

He leans heavily, grips my arms. "You're a Person," he says, "and I don't know what in hell to do."

My impulse is a belly-stopper of thanks. "So how come I feel so alone?" Only now do I realize how lonely I've felt. I almost cry.

He shakes his head. "What can we do?"

"I don't know."

For a minute we both stare out at the dark, then he hugs me hard. "I'll see you in there." And heads back to the hall.

—

All the next week, as the only labourer, I feel depressed, though the carpenters tease me as much as ever. When I come in from burning garbage and ask if my clothes smell of smoke, one of them says, "Yeah, you smell like you're in heat," which everyone finds hilarious. Later, when I'm pushing some of them around the gym on scaffolding, the lights suddenly go out and a male voice yells, "Kate, let go of me!" There's a powerful camaraderie among them, and I think they like me too, but that sense of fun, of ease, always translates around me into sex.

Earlier this summer I rescued a fluffy white kitten I've named Mr. Moon, and these days I'm feeling like some weird chick living in the woods with her animals: fascinating, but don't touch. It has something to do with working at the school, something to do with the fact that I have several lovers. Tim has come over a few times from the city but now has left for Maine. I occasionally see Mike, and there are others. A male friend once told me that the island men have agreed: no man would want to sleep with me because, he explained, "You're too strong. You'd crush them." I'd laughed, then. But later, when he and I are in bed for the first time, a wasp stings him at the vital moment and he takes it as a sign of something malevolent.

"You have wasps in your bed!" he accuses.

"Sorry..."

He's already groping for his clothes, rushing to get away. I try to joke: it's the middle of the night, I didn't put it there, I have a medical kit. But he doesn't look at me. As if I'm a witch, as if I have some kind of power that commands wasps to bite. He's downstairs, shirt half-buttoned, shoes in hand, running out the door. Ridiculous, I tell myself, to be that afraid of a woman with muscles. But this man defied the others, and now his fears are confirmed. He'll surely tell them.

I spend the next day alone, cleaning the cabin, making bread, and at 5:30 Valerie knocks at the door.

"I've brought supper," she says. Her whole face opens when she smiles.

Together, we watch the sunset, a soft pink and blue that turns vivid orange before it heavies out to black. We hear an odd blowing sound and watch as, out on the water, a huge black fin is followed by a sleek body, then another. They rise and blow and dive again, over and over, moving fast toward the south, two of them, orca whales. It feels like a gift, the universe saying, "It's all right. Really. Here's what's important."

—

At the official opening of the Island Community Hall and School, the gym is crowded with people admiring our work. People I've never seen before say how beautiful the new building is, and I feel gloriously, personally responsible for this marvel. I feel as if I've just had a baby. And there's more. Ted, the carpenter I've most enjoyed working with, has asked me to work for him as a carpenter's helper. It's a promotion, a job that calls for a little more skill and pays a bit more, too. We agree I'll work part-time, and the rest of the time I'll finish, by long distance, the two courses that remain for my master's degree. Then there'll be just the thesis to write.

Ted asks me to get a phone so he can call me on the days he needs me, but after almost two years without one, it feels weird to have a disembodied voice in my house. I keep asking the first caller, "Where are you? Really, where are you?"

Ted's company is called Salish Construction. I hear through the grapevine that one of his partners is not happy Ted's hired a woman. When I talk to the unhappy one the next time I see him, in the bar, he pushes himself far back in his chair, crosses his arms over his chest and says, "I won't be working with you." And he never does.

On the morning I start, Ted says, "We're building forms." I don't know what a form is but I've learned enough to keep my mouth shut. I'll find out. The men—three of them—build, while I bring materials: nails, two-by-four, ply. A form, I learn as I watch it rise, is like a jelly mould, only instead of adding sugar and colour to water, you add sand and gravel, mixing it before you pour it into the form. The Jell-O that sticks it together is cement. The trick to building a form, Ted tells me, is to build in negative: what juts in on the form will jut out on the final product. You have to build it strong, too, he says, so the walls won't bulge or break under the pressure of the wet concrete.

On the morning of the pour, there's a strange tension in the air. This is a big job, so instead of mixing it by hand as he usually does, Ted has ordered pre-mixed concrete from the batch plant on Salt Spring Island. As we wait for the cement mixer to arrive, carpenters flutter around the walls—a nail here, a brace there—and Ted calls me over to where he's putting together some kind of wooden shield with a brace on the back.

"Backboard," he says, and asks me to build another one. My first solo project! That morning he'd shown me how to "toenail"—drive the nail at an angle—but now that everything's in a hurry, I can't seem to hammer. When I finally get it done, I'm amazed at myself.

Ted calls for an early coffee break, but the instant we hear a groaning on the road, the carpenters are up and ready. The truck is huge. I've noticed these strange vehicles in the city, looking like monster martini mixers. All the concrete in that truck will have to be moved, one tiny wheelbarrow at a time, into the forms.

The driver lowers a slide and releases a stony soup that rattles into the first wheelbarrow lined up to catch it. My job will be to hold the backboard that directs the concrete in between the form walls. I put all my weight behind it but the first time the mix thuds against my backboard, it's as if a sumo wrestler has hurled himself against me.

"Hold it!" Ted yells as he struggles to keep the grey mix pouring into the form. He's barely finished when someone else yells, "I've lost half a load! Get that backboard over here!"

I race along the platform to where the man is swearing, fighting to keep his wheelbarrow upright. I barely manage to put my shoulder to the board before the load hits and rattles down into the wall.

There are three of them and someone is always yelling. There's no gentle in their voices, no "please," and my feelings are hurt until I get caught up in the excitement.

"Puddle!" Ted orders and hands me a long, thin stick. "Up and down!" When I stand there, confused, he takes the stick out of my hand, not unkindly, and jabs it repeatedly into the freshly poured mix.

"As deep as you can get it!"

"Why?"

"Takes out bubbles," and he's gone. For a few minutes I puddle, the motion strangely sexual, then, "Backboard!" someone yells.

I like it. I like the rush of concrete, of using all my strength to make the mix go where we want it to. After a while, I begin to know what's coming, who'll need the backboard next, and they hardly have to yell for me at all. Whenever there's a pause, I puddle.

When the truck is empty and the carpenters are putting final touches to the walls, one last wheelbarrow full of concrete waits to be placed. I pick up the handles, knowing to let my arms go long and stiff so my legs can take the load. Wheeling wet concrete is like steering a huge pan full of wet dough into the oven. Only heavier. I'm almost at the wall—I can do this!—when the wheelbarrow takes an alarming swing to the right. I throw my weight left to counter it and succeed—too well. The load shifts wildly leftward, and to my horror two-thirds of it slops to the ground. That morning I'd proudly put on the T-shirt my family gave me last Christmas. "The best man for the job," it announces in bold white type, "may be a woman." I cross my arms over my chest. When I look up, Ted is grinning.

"We were saving that load for you," he says. "Too full for the rest of us." Then, amazingly, he ignores the spilled concrete. "Bang the forms with your hammer!" he orders, and hits the nearest wall to illustrate.

I'm confused. Haven't I just made a big mistake, maybe an expensive one? Yet nobody's mad. Nobody, at least for now, is going to fire me. My right knuckles are swollen and blue from a knock I gave them during the pour. My right wrist throbs, and when I go home I'll put an elastic bandage on my right arm from the pain of so much hammering. But I feel fine. I have built a wall.

On Monday morning, when Ted says, "We're going to strip the walls," I cringe. "Strip," he'd said, but no one snickers. I don't ask "What's stripping?" I watch the men out of the corner of my eye and allow my body to shadow Ted's as he tears down the lumber we put up the week before. When he lifts an arm to bring his hammer down hard against a brace, I let my hammer fall the same way. His hammer is faster but I push to keep up. I don't think; I mime construction. When only the sides of the forms are left, Ted lays his bare hands on the top row of shiplap and pulls hard. The lumber falls, and without a pause he moves to the next. Like his puppet, I go to the opposite side of the wall and pull. Nothing. I pull harder. There's a faint sucking sound but the board stays put. Ted's already halfway down the other side. This time, rather than a steady, polite pull, I jerk with all my strength. The shiplap surrenders with a slow-motion ripping noise and comes away slowly in my hands. Beneath it, like sculpture, the wall is a smooth dark stone that shows a few tiny bubbles and the imprint, like a negative, of wood grain. It's beautiful.

All morning I stand in a muddy ditch and throw my strength into ripping boards off concrete with my bare hands. The product is beautiful, but there's something alarming about the process. In the planer mill, and on scaffolding at the school, I'd learned how to move my body in big, exaggerated motions as if swimming through air. But the movements of stripping concrete are even bigger, force me to open wider my arms and legs. It's almost rude. Definitely unladylike.

Maybe that's why, next time I go to the city, I end up in the lingerie section of my favourite second-hand store. I don't know what I want until I see it—an old-fashioned silk nightgown that hugs me all the way down, from the slightly tattered beige lace at the shoulders, across my breasts and belly to my ankles. I never wear it for lovers, only when I'm alone, after my bath. I can float around the cabin and out onto the deck feeling elegant and, above all, feminine.

One evening after dinner, as I sit admiring the peach sunset outside my window, my eyes fall on the jar of flowers Jeanne has brought me, sitting on the table in front of me. My dinner had included pickled beets, and when I spot a drop of the deep red juice splashed on a yellow dahlia petal, I begin to cry. I have no name for the sharp voice that increasingly underlines each day. Who do you think you are? it whispers. What kind of a woman does this work, anyway?

Yet I love the work. I love its sense of order: First comes A. You can't build a roof without a foundation, so B must come second and only then can we build C. Last week we poured and stripped the foundation.

On Monday we put the joists down and, over that, a plywood floor, and for a few days the carpenters have been building a sort of wooden box for which I bring endless two-by-fours and ply and nails. When we lift it, we've built a wooden wall. Ted laughs when I want to take a picture.

I'm drinking too much, I know, but I go to the bar three and four times a week now, for company. One night I get into a discussion with two of the carpenters from the school about the gym ceiling. One of their wives says irritably, "There are the men, talking shop again."

And the crevice that has been deepening inside me gaps open. I've been caught up in the exciting talk about building, yet I'm also interested in what the women have to say. Where exactly do I belong? All the definitions have turned upside down and I'm afraid I'm going crazy again.

Then I meet Judy.

Ray Hill (from the school) and his wife, Beth, have become friends. I'm at the end of a weekend visit to their place on Salt Spring Island and Ray is honking at me to hurry so we don't miss the ferry when another woman arrives. She's about my age, tall with brown hair to her shoulders and an open face with alert eyes. I like her immediately.

"You have to meet our friend Judy Currelly," Beth says. "She works as a bush pilot."

"Bush pilot! I'm a construction labourer!"

Ray honks again. "The men," I say. This is urgent. "How are the men?"

"Mostly fine." Judy tells me it's only the odd man who gets as far as the door of her plane before he sees it's a woman pilot and refuses to fly with her. And then?

"I fly without him."

"If you don't come right now...!" Ray's yelling.

Judy and I shout phone numbers at each other, and all the way home I pace the ferry deck. It isn't any particular thing she's said; it's my vast relief at meeting her. I'm not alone.

By the next morning I know what I must do. Before I can go further in construction, I have to find out two things: first, if there are more women like Judy and me, doing what they call men's work, and second, what it's like for them. I'll finish my degree and this will be the subject of my thesis. But I'll work on the island for just a while longer, both for the money and because I like working with Ted. He's patient, even when I'm at my most awkward—like the time he held a stake for me so I could try the sledgehammer. I'd missed the stake entirely, hitting his shin, and when I asked if he was okay, he'd gasped, "Fine," before, jaw clenched, he'd reeled away, clutching at his leg.

One day when it's just the two of us starting a new job, I park behind Ted's truck at a newly cleared space on a barely developed road. In the centre of the lot is a deep hole fringed with evergreen and a few thin alders. I leave the truck door open, as always, so Ruby can climb in and out.

Ted gets me to hold a white stick that he peers at through one-eyed binoculars—"builder's level," he calls it—then shows me how to use it. When we take a break, sitting on a pile of dirt to drink our tea and throw sticks for Ruby, he says, "Ever thought of an apprenticeship?"

I've never heard the word.

"That's how you get to really know how to do all this stuff. You work four years for a carpenter who teaches you on the job, and for six weeks every year you go to school to learn theory, like how to read a level, or cut rafters."

Not for a second has it ever crossed my mind to become a carpenter—I'd never dare. But I like this work better than anything else I've ever done, and I'm thirty years old. One of these days I'm going to have to decide what I want to be when I grow up. When I go to his place that Saturday, Ted shows me four binders, one from each year of his own apprenticeship training, each one full of tools, concrete forms, stairs, doors, rafters and numbers—lots of numbers.

"How's your math?"

Math, especially geometry, had been my favourite subject in high school, until the guidance counsellor told me girls didn't need math and I'd dropped it. Ted tells me it will be easy to pick up again, though I'm not so sure.

"You could go straight into apprenticeship," he continues, "or you could take what they call a pre-apprentice course first." He gives me the address of a trade school in Victoria. "You'll want to apply for construction carpentry," he says as I'm leaving, "not cabinetmaking." The difference is that cabinetmakers make furniture, and construction carpenters build houses.

As I drive back to my cabin, I'm thinking, why not? I want to keep building things, and what he's shown me looks a lot more interesting than labouring. Besides, I'm not sure what I can do with a master's degree—when I finally get it—and in the meantime I can work in construction, making good money.

"And I'm not too bad at it, either," I say aloud to Ruby. I would never tell anyone except my dog, but I know this is true. Ted wouldn't have told me about apprenticeship if it weren't. So, what the heck, I apply, and in May I get a phone call from the school accepting me to the upcoming Pre-Apprentice Program in construction carpentry. I ask the man so many questions that he gets the impression I'm asking (and maybe I am) if I got accepted just because I'm a woman.

"You were put on the list along with all the men who applied," he says, "and we've taken you into the class strictly on merit."

I don't tell him about the master's degree. I tell him I want to spend time with my family over the summer. He says I'll have to reapply for the fall, but that shouldn't be a problem.

I still have several months left on the lease for the cabin, but I rent a room in Vancouver as well, so I can also be closer to school. I'll finish my degree and then...? I'm not sure what then.

6

contradictions

It takes a while to find a professor who'll agree to be my advisor and shepherd me through a thesis on women who do "men's work." I think the problem is that no one's done much academic research on blue collar male workers, let alone blue collar women. I know Liora Salter from taking an earlier class with her; as a teacher, she's clear, well-organized and smart. "Non-traditional," she says, is the new name for women doing work usually done by men, and since almost no one has written about them—us—most of my research will be what Liora calls "primary," interviewing the women and asking what they think. First, there's something else I must deal with.

I'm pregnant, and the question is: do I want to keep this baby? I'm thirty years old, prime. Over and over, my head says no, but my body is screaming for it. The father has refused to be involved but I'm clear—abortion won't be possible after three months and I'm not going to passively let time make this decision for me; I have to decide by the end of June.

On June 21, there's an outdoor solstice dance on the island. After a toke or two, I know that the answer to what I must do is here, under the full moon. I let the question go, stop thinking, and dance. The field where the party is held is in a long valley with low hills on each side. I'm surrounded by friends. There's music and talk and laughter and love, and tears begin to seep from my eyes. I ignore them, dance until tears pour down my cheeks, and I know I will not have this baby. I am utterly certain it's the right decision, but there's grief, too, so I keep dancing.

I dance for the child that won't be mine, dance to send its spirit on to some other body that will be more welcoming, grieve for the child and for myself who will not experience childbirth—and feel profound relief. Jeanne, one of the only people who knows, takes one look at me and knows, too. We hug and sway together for a long time.

Later that summer, Liora invites me to come to a conference she's hosting on Harold Innis, a Canadian economist I've barely heard of, who's her main academic interest. I turn up—not very enthusiastically—wearing my favourite second-hand burgundy men's shirt with big loopy white flowers and a long burgundy skirt I sewed from fabric on the dollar table. I'm glad I wore my best clothes when I notice one of the men on the afternoon panel, a grad student, like me, but in Political Science. I ask a question so he'll notice me, too, and at the reception after, I go to where he's talking with Liora so she can introduce us.

His name is John. He's tall and slim, sternly handsome, with beautiful bones—high cheeks and a strong jaw—and sweet, vulnerable eyes. He's quiet, but when he speaks, I like the careful way he chooses his words, what good sense he makes. We chat, we say goodbye, but on the island the following Saturday night at the pub, I notice him leaning back on his elbows against the bar. Is it coincidence, or did I tell him where I live?

"You're John Steeves."

He remembers me, too. I invite him back to my table, pull up a chair beside me bold as brass, and, when the bar closes at eleven, offer to drive him to the after-party.

I watch him carefully: how he doesn't seem alarmed by a woman with a truck, how easily he climbs in, how kind he is to my dog. At the party he talks mostly to other people he knows—I'm surprised to find he used to live here too—but toward the end of the evening he gravitates back to me. I drive him to where he's staying and it's because I'm interested in him that I say goodnight with a wave, not a kiss. As I drive away, I'm smiling because I know he wouldn't have minded, because he didn't push it with even one last look, because we've exchanged phone numbers.

The next time I'm in town, John invites me for coffee. We meet in the university pub.

"How's it going?"

"Fine." Silence.

"Classes good?"

"Yes." Silence.

I chatter for the minimum time required by politeness, excuse myself and leave, figuring that's one guy who'll never call again. Clearly, I've bored him to tears; I've bored myself. But he does call. He suggests a movie, which means I don't have to talk the whole time, but after, over coffee, things start out the same way: he speaks in monotones, I chatter. After ten minutes of this, I'm pissed off. Why am I doing all the work? I shut up. If he's not going to talk, then neither will I. I look around, wonder if there's anyone else here I might know. Ten seconds, fifteen; it feels like a full minute before he says something, and what he says is actually very interesting. Picking my words carefully, I reply. He replies back, and slowly, as if a small, precious flame has begun to burn, we have a real conversation.

My way of relating to men has always been with long, intense talk and quick sex, or not. John isn't into long, intense conversations, though he's very smart, and when he does speak he goes like a magnet to the core of things. One day after we've driven somewhere and I've chattered for ten minutes, first to say what's on my mind, then to fill the looming silence, I decide I'll outwait him. When I can't stand the silence any longer I say, "What are you thinking about?"

He glances over, a little surprised. "Nothing."

"You're thinking about nothing," I repeat, disbelieving. "Really?"

"Yeah, really. I had nothing in my mind."

I'm quiet for a while, trying to absorb this amazing fact. It must be a guy thing. I'm always thinking, mostly about what the guy might be thinking.

John says, "What are you smiling about?" and we talk.

Soon we're seeing each other every time I'm in town, and still we haven't even kissed. Then he invites me to his communal house for dinner. He lives with two other couples, he tells me, their daughters, and John's son.

Son?

From a previous marriage, he has a seven-year-old, Kevin, who lives with his mother except for some weekends. Tonight, Kevin's with friends.

One of the women in the house is a First Nations activist and her husband is a lawyer. The other couple are Jude, a daycare worker, and her common-law partner, a biologist. I like it that John lives in a house where politics matter, where the politics are Left. As I'm leaving, he asks me to stay the night. I've considered it but say no. This feels too serious to fool around with.

"Are you sure?" He has beautiful bones.

"Thanks anyway." But I can't resist reaching out to touch his arm, and there's a bolt of energy where we connect.

"A walk then?" he asks, and we walk. And have our first argument. It's over the way I told his friends about an upcoming rally for abortion on demand.

"You didn't have to fling it in their faces." He's so calm.

But I'm furious. Here it comes, another man telling me what to do and how to do it. I did have to throw it in their faces—it was a test and they met it, so what's the problem? As we work our way toward making up, I assume I will now—as I've done with every man I've ever known—prostrate myself to his point of view. But when I do that, he ignores me.

He doesn't want a doormat! I'm dazed. Psychically, I stand up and dust myself off and, for perhaps the first time ever, look at a man as a partner, eye to eye. The next time John comes to the island we meet again at the bar, but this time he comes home with me after, and stays. He doesn't mind that my hands are rough or my body very strong. The next time I'm in town I stay with him, and in the morning, as we come downstairs for breakfast, there's a pounding on the front porch, the noise of children. A girl of about ten and a boy a bit younger explode through the front door. The girl runs on into the kitchen, but the boy stops suddenly and looks up. He's tall, solidly built with a round face, sad brown eyes, sandy hair and an aura of watchfulness.

John says, "Kate, meet my son, Kevin."

The child scrutinizes me carefully. "Here we go," he says, like some small oracle, then chases after the other child.

I've never wanted children, but I'm curious, and John's and my dates begin to include Kevin. I like being with the two of them, their quiet, sober company. Once when the three of us rent a boat to go fishing, I'm impressed when Kevin catches the first fish, then a second.

"How do you do that?" I ask.

"I think like a fish," Kevin says solemnly as he drops the baited hook back in the water.

Perhaps what moves me most is John's care for his son. The second or third time I sleep at John's house, Kevin wakes us at three in the morning because he can't sleep. Without even a moan, John rolls out of bed and makes the boy a small mattress on the floor—foamy, sheet, blanket—and we all go back to sleep. After the first few times, I wait for John to say it's time for the woman to take over the childcare, but he never does. It feels safe to love a man who has this much compassion for his child. It feels natural to stop sleeping with other men. When I want comfort now, I want John. Increasingly, I spend all my spare time in the city with him and Kevin.

Meanwhile, I've been phoning all over town looking for women who've worked in pulp mills, saw mills, mines, or on fish boats. When I don't find a single one in Vancouver, Liora advises me to head north, but before I leave, John asks to come over for coffee. When he arrives I'm sitting out on the back step, trying to catch some air on one of Vancouver's few really hot summer days. He stands a few steps below, facing me.

"Kevin will be starting a new school this fall," he says after a long silence, "and will be living with me from now on, so I'll need a new place."

It's my turn to be silent.

"What do you think about us all moving in together?"

I hadn't thought of it. I like the little guy, but we're cautious with each other, aware we share a man we both care for.

John's face is more open than usual, and worried. He wants me to say yes.

"Sure," I say. "Why not?" If it doesn't work out, I can always move out again.

The three of us celebrate our new family with a camping trip to the Queen Charlotte Islands (later, Haida Gwaii) off the north coast of BC. Then John and Kevin go back to Vancouver to find a house for us and I stay in the north to look for non-traditional working women.

I have one name—a logger on the Queen Charlottes. She lives alone in the woods without a telephone so I can't call ahead, just tromp through thick forest until I spot a small log cabin where I knock, hoping she's home. She's gracious, offers me a seat at her table, a cup of tea, then answers the questions I pepper her with: How did you start? What's it like working with all men? Has it affected your femininity? What keeps you here?

She finishes by giving me one other name, of a woman blaster in Prince Rupert. Blasters set off dynamite charges for road building and rock clearing. The blaster likewise welcomes me, and she gives me the name of one more woman. I can't afford motels or restaurants, but the women seem to sense this. Over and over, when I tell them what I'm doing, why I'm here, our conversations go something like this: "Have you eaten? Need a place to stay? Stay here, eat with us, then you must talk to A." A sends me to B and so on until, after three weeks, I've talked to twenty-two women who work as fishers, pulp mill, mine and sawmill workers, mechanics, labourers and a single carpenter. Their extraordinary generosity—sisterhood—makes my journey, and my thesis, possible.

The details of each story are different, but each woman's love for the work is as strong as her determination to keep doing it. I love it, they tell me; I hate it. I feel more feminine; less. The men are wonderful; the men are awful. The work is easy; the work is hard. I step off the Greyhound bus in Vancouver with a bag full of precious audio tapes and a book thick with notes.

John has found us a house to rent in Vancouver's East End. The Saturday after I get back, we drive to the Sally Ann and load the back of my truck with a red and green tartan couch and arm chair, a wooden table and four kitchen chairs. We buy a gallon of salmon-pink paint for the kitchen and another that's supposed to be coffee colour but that our landlord disapprovingly calls "Nightclub Orange," for the bedroom.

Everyone's busy—Kevin in school, John as a labourer and fitter at the shipyard (until he can find a better job), and me writing my thesis. Routine is balm, and I'm so happy to be with this man who quietly goes about making us comfortable. It's John who thinks to put a light over the kitchen counter so we can see what we're cutting, who builds an extension to the counter so there's more work space. We divide all costs and all chores—cooking, cleaning, shopping—scrupulously in half. Even Kevin takes his turn cleaning and, later, also cooking one night a week, with John or me to help. I live on my teaching assistant's salary at school and a minimal student loan.

One night as I'm reading to Kevin before bed, I put my arm around him and feel a sudden catch in my throat. I'm his stepmother, though all I know about stepmothers is that they're unambiguously evil in fairy tales. When I check the library, there's exactly one book about stepmothers and it's fiction. What does a stepmother do? None of us even knows what I should be called, but because Kevin spends every second weekend with his birth mum, who lives nearby, we agree on Kate.

When we're alone together, Kevin keeps up anxious conversation.

"Are you having tea?"

"Um huh."

"What kind of tea?"

"Black."

And as I add milk, "You take milk?"

It almost makes me laugh, but I feel for the child's fear, his worry. I worry too. When it's my turn to cook, Kevin helps me peel potatoes and set the table. How can a seven-year-old be kind? One day when John's away, Kevin comes home early from school because "I got in a fight." I mutter something about how "fighting's not good," which sounds lame, even to me. What would a real parent say? Kevin listens, then says patiently, "Kate, little guys pick on me because I'm the biggest in my class. So I have to fight. It's what boys do."

I'm speechless in the face of the startling good sense of it. "Right," I say, "I see." And I do.

I've set up a desk and file cabinet in one of the bedrooms and try to stick to a daily schedule, transcribing interviews and reading. But it's lonely. At Liora's suggestion I submit a proposal to the Canadian Research Institute for the Advancement of Women, and later to the newly formed BC Human Rights Commission, to write reports on women in non-traditional jobs in the province, based on my thesis research. When both accept, I'm scared shitless. I have no idea what "research" is, or where all this is heading. All I have is the small voice in my gut that tells me that what I'm doing feels right. Still following my gut, I apply again for the Pre-Apprentice program in Construction Carpentry, this time to a Vancouver school called Pacific Vocational Institute.

—

Since John introduced us at his communal house, Jude and I have become good friends. If there's a women's event—a Take Back the Night march or a chance to look at our vaginas at the Women's Health Centre or a Women's Day march—I call Jude. She's warm and funny and wise and feels like a sister in every sense. On October 14, 1979, we go to a conference on Working Women organized by the BC Federation of Women, Working Women Unite, and Langara, a local college. Jude heads to a workshop on women's health and I go to one on women in non-traditional work, fearing there'll be exactly two of us there. A few weeks before, I'd had a phone call from a woman who was looking for ideas for workshops.

"What about one for women in non-traditional work?" I'd suggested. "Would anyone come?"

"I'd come," she said. She's a printer, learning her trade at a shop called Press Gang that's run by women.

When I arrive, the room is packed with over thirty women. I double-check the door number, but sure enough, there's my printer, grinning from the far end of the table. We start by going around the table to introduce ourselves.

"I'm Judy," the first woman states, leaning forward so she can see each of our faces. She's beaming, forearms planted firmly on the table in front of her. "And I'm a carpenter." She says it as an exclamation, which it surely is, and with such joy that we all laugh.

When we've gone around the circle—carpenters, shipwrights, printers, a welder, bus and truck drivers, a seawoman, electricians, electronic technicians, mechanics and mill workers, among others—the room erupts in excited talk. The printer tries to keep a rough order as we interrupt each other, eager to hear all the details: How did you get your job? What about physical strength? How are the men? We laugh at the questions. We laugh at the answers. We laugh at the joy of being together, and at the end of an hour and a half, there's no question we'll meet again, form an organization so we can keep in touch. When we do meet and it comes to naming ourselves, one woman suggests, "Women in Non-Traditional Work."

"The men don't call themselves Men in Traditional Work," another remarks dryly, and that decides it. We're Women in Trades—WIT for short. We meet once a month, to talk. It's such a relief to have others roll their eyes at precisely the right moment in your story because they know, they know exactly, how it feels.

"The pressure at work is like a ton of feathers," Janet, an avionics technician, says at one of our first meetings. "The first time someone does a double take when you pick up the drill doesn't bother you, or the hundredth. It's the thousandth time that crushes you."

Alice, a motorcycle mechanic, also teaches a self-defence course for women called Wen-doh that she agrees to teach to Women in Trades. She starts with showing us how to use our strongest body parts against an attacker's weakest: elbow to rib, heel of hand to nose, knee to crotch. At the end of the course, she's promised, we're each going to break a board with our bare hands. Sure enough, on the last day, she places fresh pieces of pine across two grey stone blocks, one after another, and we take turns kneeling in front. When I see the large knot in the middle of my board, I have a moment of panic, but Alice is calm, standing beside me, arms crossed.

"You can do it."

I focus, centre my attention as she's taught us, lift my hand and BANG! The board falls into two parts, splitting neatly around the knot. I keep it in my office where I can see it whenever I feel discouraged, which these days is often.

For weeks I've been sitting on the floor of my office surrounded by a daisy chain of paper, trying to make sense of why the women I interviewed contradict themselves over and over. "I love this work," they say, and, a few minutes later, "I hate it!" When I tell Liora the interviews are full of contradictions, she says, "Maybe that's the point."

The problem, I figure out, is that men and women on the job are caught in double binds. A double bind says that no matter what you do, you lose. When the first woman comes on a job, she works particularly hard to prove women can do it; and the men resent her for showing them up. If she doesn't work particularly hard, it proves women can't do this. Or the men stop swearing "out of respect for the lady," then resent her for making them change. If, to make them feel better, she starts to swear, they're shocked. Or the men act as "gentlemen"—the only role they know—and carry her lumber or her welding tank for her, then complain she's not doing her job.

Now I have a conclusion for my thesis: the old roles don't fit; the double binds prove it. When women do the same work for the same pay as men, it's no longer appropriate for her to play "lady," nor the man, "gentleman." If being "feminine" means being soft, weak and mechanically incompetent, then we need a new definition of feminine. In fact, if "masculine" means taking the tool out of the little lady's hand and doing it for her, then we need a new definition of masculine, too.

In December I pass my oral exams and hand in my thesis at SFU, and in January 1980—with a master's degree fresh in my back pocket and a bad case of nerves—I turn up for my first day in pre-apprentice construction at the Pacific Vocational Institute.

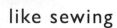

like sewing

Jude has a small piece of property and wants one day to build herself a house on it, so she's signed up for the pre-apprentice course with me. As we walk into the building marked Carpentry Classroom, it's a relief to have not just another woman but a good friend close beside me. The room is a cheery yellow. Underneath a row of windows on the far wall is a collection of what look like dolls' houses. To our left is a chalkboard and a teacher's desk and to the right, several rows of wooden tables and chairs filled by men. As Jude and I sit down in the two empty seats, my cheeks grow hot.

At 8:00 a.m. precisely, a tall, heavy-set man walks in the door wearing a blue lab coat with "Mr. Stewart" written on the white patch over his left breast pocket. He gives Jude and me the barest nod. He talks about the course until coffee break, and after, we do math until lunchtime. After lunch we follow him outside, across a service road and through blue double doors into a metal-sided building he calls "the shop," big as a football field, two stories high with metal walls and a concrete floor. To our left are table saws and power tools, with work benches at the far end. In the open space between them, a clatter of young men hammer and saw. Past the benches, a friendly-looking man with red hair and a bushy beard leans over the top half of a bright orange double door, watching us approach.

"This is Bob," Mr. Stewart says as we gather around. "If you want a tool, you come here to the tool crib and ask him." Mr. Stewart has us form pairs (Jude and I make instant eye contact) and check out a builder's level, tripod and rod—the same tools Ted had shown me on the island that led to my being here. It's a good omen, so I'm not sure why I'm trembling. Outside, Mr. Stewart shows us how to set up the tripod and attach and level the instrument by spinning the three knobs at its base.

"Always move your thumbs in the same direction." He fiddles for a second until the tiny levelling bubble hovers precisely between two black lines. Now it's our turn. Jude goes first. Every time she turns a knob, the bubble swings wildly, and I do no better. Our fellow students have all levelled their instruments and are reading elevations when, in desperation, I kick one of the legs, which makes things worse.

"Move along," Mr. Stewart tells us finally. "You can practise later." We pack up our instrument and follow him back to the shop where everybody crowds around a cart full of tools.

"Can you name these?" he asks the group.

I know framing square, tin snip, crescent wrench (that lollipop with a mouth). But I don't know rip claw hammer. Others I recognize but had forgotten until one of the guys calls it out: nail set! plumb bob! line level! Jude and I fall back as the men press forward, eager to call out names: combination square! slip pliers!

On the way home that afternoon, I don't ask Jude how she's feeling about our adventure, and she doesn't offer.

There are two bathrooms in the shop, both labelled Men, and the next day Jude and I ask Mr. Stewart to make one of them a women's.

"They're men's washrooms. Always have been," he says. So every time one of us wants to use a toilet, we have to run through the shop, across the yard, into the classroom building and back.

Then there are the boots: it seems they don't make steel-toed boots for women. I'm lucky my feet are big so that with a pair of heavy socks I can get along with the smallest men's size. But Jude has small feet, so she asks Mr. Stewart if she can wear hiking boots in the shop. He says absolutely not—for safety reasons—so she stuffs men's boots with three pairs of socks and still skates around inside.

In class, the one with a hand up to ask a question is always Jude or me.

"Yes, girls?" Mr. Stewart says, until Jude asks him to please call us "women."

One question in particular we ask, over and over: "Why?"

"The proportions of a right-angled triangle," Mr. Stewart tells us one morning, "are 3:4:5."

Our hands shoot up together.

"Yes, girls?"

"Why?" Why does a foundation have to be perfectly square? Why do you have to set the foundation below frost line? Why do you need a nut with a washer? I worry that I'm stupid. How do the men already know all this stuff? And even if I am stupid, if I don't ask my questions here, I know the carpenters won't be eager to answer them when we're out on the job.

Every morning in class we do theory, and every afternoon in shop we learn the basic practices of construction. We sharpen chisels in a dramatic blaze of sparks, whet them on a sharpening stone, cut and glue a sanding block. After we've practised using a handsaw, soft-focusing our eyes on its back to keep the blade straight, we make a float "for rough-finishing concrete," Mr. Stewart says. The wood is cedar: sweet smelling and so soft that our chisels gouge and stick. Jude jokes with me after that it's like a lover: "Fascinating, so I can't stay away, and infuriating because it won't do what I want it to."

The power tool lesson comes with a heavy dose of safety: Wear goggles! Don't stand behind the saw! Where are your fingers? Finally, we begin to build real things, starting with a tool box with a round dowel for a handle, though the dowel is much too big for Jude's and my hands until Bob, in the tool room, shows us how to rasp it down.

Jude and I are the clumsiest ones in shop and have to work harder to tighten things and loosen things, so we start going to the gym twice a week to lift weights. Still, we're always the last ones finished, and every night on the way home we console each other that tomorrow, for sure, we'll catch up.

—

"The basis of female oppression is force."

I've come to hear Kate Millett, the author of *Sexual Politics*, give a lecture to a room packed with women. She says, "Another crucial element in our oppression is shame over our female sexuality," and I nod—yes. All those men, so uneasy about my sleeping with other men, even while they slept with other women.

"All women are sisters," Millett finishes. "We're having the time of our lives. Remember that. We have the immense thrill of changing our own lives, as women."

Discussions and meetings like this are my lifeline.

⏤

There are fourteen men in our pre-apprentice class. I can imagine working with a couple of them, especially Dominic, an Italian guy with a crush on Jude who works with his dad in construction, and Tony, who has trouble reading. Every morning Tony asks me, casual-like, what our homework from the night before was about. But he has no problem in shop; in fact, he's always the first to finish.

Mr. Stewart is hardly ever in the shop with us because he's the chief instructor and has a lot of meetings. When we have questions, we ask Bob or Dominic. On the first day of school, Mr. Stewart promised that after we'd each built a picnic table, the group could build a real house; but everybody's already on their second table, and when Dominic is told to build a third, there's grumbling.

"Why don't we complain?" I ask the others at lunch.

"Who to? He's head of the department."

"We'll tell him we know how to build picnic tables and now we want to build the house."

Someone snorts. "Great. We've got ourselves a union." But we all agree, and after lunch I find myself walking with two of the men through the shop and up to the instructors' office. Mr. Stewart looks ruffled when we tell him we want to get on with the house, but he says he'll see what he can do, and the following Monday it isn't Mr. Stewart who greets our class, but a small, wiry man who introduces himself as Art Green. Everyone perks up.

"I'll be handling this class as if I'm the foreman and you're all carpenters," he says. "Starting today, we're going to build a house." We pull out our pencils for a lesson on concrete footings.

"Call me Art," he says that afternoon when one of us calls him Mr. Green. "No one calls anyone 'Mr.' on any job I've been on."

A whole new language begins to roll off my tongue as night after night I memorize anchor bolt, strap tie, snap tie, Easy-Strip, pier. That Friday, Art divides us into two crews—Jude on one, me on the other. Monday morning, he says, we start building two real houses.

⏤

At home, I treasure our routines. Every Saturday morning, after breakfast, the three of us climb into John's battered green Volvo to do chores: groceries and credit union and our weekly check-in at the second-hand store. Still, I don't get this "stepmother" business. One day when

Kevin and his best friend, Eddy, burst through the front door, running past me toward the back, I shout, "Don't run in the house!" I don't really mind, but my mother used to say, "Don't run in the house," so I say it, too.

Kevin stops dead in his tracks and gives me a look of—well—pity.

"Okay," he says, as if speaking to a slightly younger child. And with great decorum he follows Ed, who'd hightailed it as soon as I opened my mouth.

I feel like an idiot. To tell the truth, I'm a little afraid of Kevin, afraid that if I can't work it out between us, it will be seriously bad for my relationship with his dad, which isn't going all that well. John's job at the shipyard doesn't seem to be working out, but he won't talk about it. In fact, he doesn't seem to have much to say to me at all. I try to please him but I never seem to get it right, and we start to niggle over small things. Instead, he focuses on Kevin.

Kevin isn't doing well in grade four, which is maybe why he keeps getting into fights. One evening after supper, John suggests we all go to the local library, where he and Kevin disappear into the kids' section and emerge with a comic book about a chubby Gaul named Asterix. From then on, we spend every Thursday night at the library, where John and I browse while Kevin picks out a new Asterix, which leads to a general fascination with all things Roman and Gaulish. When I start to call Kevin "Kevinex," he doesn't object.

One evening when the child announces he's bored, John says, "If you're bored, it's a failure of your imagination"—this is why I love this man—and for the next few weeks, the two of them spend their spare time making a heavy wooden shield with a red and white design like Asterix's, then a sword and, to complete the outfit, a paper helmet.

John and his ex-wife, Kevin's birth mother, have made it very clear I am not Kevin's mother. Then I overhear John telling Kevin's mum, "He was in the principal's office again today, second time this week."

When John gets off the phone I ask why he hasn't told me, but he turns away as if I'm not worth answering. Suddenly, I've had enough. I may not be Kevin's born mother, but I'm sure as hell acting like one.

"I have a right to at least know what's going on with him!" I snap, and stomp out of the room. After that, John includes me a little more in decisions about Kevin.

It doesn't help that John's increasingly desperate to get out of the shipyard. Every night he comes home and, if it's not his night to cook, lies on the couch until supper, then goes back there again after, not talking.

Sometimes he throws a tennis ball against the wall, endlessly thumping, driving Ruby crazy as she tries to catch it. Which drives me crazy.

"She loves it," he says. And sure enough, Ruby waits, panting, tail up, legs planted, eager for the next one. Which pisses me off even more. These days, what holds John and me together includes our care for Kevin. I'm starting to really like the little guy. He makes me laugh.

—

At school, I know I'm thinking like a labourer, not a carpenter, planning ahead so I can bring materials to help the others, the guys. When it comes time for ceiling joists, I don't volunteer to go up and walk the narrow two-by-fours on the top of the walls we're building. It's one thing to be on the roof of the island's school with all the space in the world for my feet—it's another to balance on three and a half inches of what Art calls "the top plates." I stay on the ground and pass up lumber.

Art, who's constantly reminding us to pay attention as we work, gives one entire lesson on the circular saw. "Above all," he says, "every time you pick up the saw with your right hand, check where all the fingers are on your left." He tells us about a guy he knew who sliced off the tops of three fingers on his left hand when—just for a moment—he made a cut without paying attention. After that, I look for my left-hand fingers every time I use the power saw, and once, I move them.

Jude and I are better than most of the guys at theory. It's fun figuring out quantities of concrete and wood, and I get As in my tests. But there's a gap between theory and practice. In theory I know exactly how to hammer a three-and-a-quarter-inch spike. It's a whole other game to actually get the damn thing in with fewer than seven or eight hits, let alone without bending it. The guys on the island used to say, "Strong wind today?" as they watched me bend nails. That's the difference between theory and practice.

When I first learned to drive, I felt I was inside a death machine, hurtling forward at warp speed, always verging on being out of control, and at the same time zinging with the excitement of motion, power and independence. Building a house is like that. I'm barely in control of tools and heights that could cause serious damage or death. At the same time, there's the thrill of strapping a heavy leather belt around my hips, the physical pleasure of sliding my hammer out of its hook with an increasingly smooth motion of my arm. There are moments now when I stop being a nervous, self-conscious watcher and become a doer, immersed in building. In those moments I am powerful.

The problem is money. When John and I moved in together, the deal was that we'd split all expenses, but I haven't brought in any income in months, and I have six more weeks to go in my course. John makes just enough to pay the rent and keep us in food, and I keep track of my half so one day I can pay him back. Still, I'm not sure how I'm going to make it to the end of this course, so one afternoon, instead of going with Jude to the gym, I go to the Student Aid office. When I tell the counsellor I'm in the pre-app carpentry program, she raises her eyebrows.

"There aren't many women, are there?"

I tell her I think we're the first, that they don't even have a bathroom for us yet.

When I explain that I'm not sure I can make it to the end of the course without some income, she tells me our course is too short to be eligible for provincial aid and she doesn't know of any scholarship for pre-apprentice courses. My heart sinks. I need a minimum of two hundred dollars to make it through to April.

"Let me look into it," she says.

A week later, she calls me in. She still hasn't found any scholarship money, she says, "but there is a one-time-only grant of two hundred dollars." And she hands me a cheque. After I get past my relief, I have the oddest feeling there was no grant, that the money likely came from an unofficial source—her pocket, perhaps.

When the walls are up and the joists almost finished, Art announces we're going to learn rafters. On the island, Ted's cutting of rafters was the greatest marvel of construction magic to me. He'd take a few measurements and mark a board on the ground that—when cut and hoisted up—was a perfect fit in the puzzle that becomes a roof. But it's harder than Ted made it look. I can do the math, but I can't translate it into wood. Theory and practice. And for a change, it isn't just a female thing. Several of the men look as blank as I feel. Over and over, with infinite patience, Art says, "Don't worry. We'll be doing this again. It will get easier."

And slowly, everyone gets it except me and one other guy. That afternoon, Art leads the two of us out into the yard and over to one of the dog houses the third-year apprentices have built to practise their rafters. He holds a piece of paper up against the real thing.

"This is my rafter." He holds the paper rafter on the flat, then raises it to the final slope of the roof. "See how it has to get longer as the rafter lifts?"

But I don't.

It's Jude who gives me the key. "It's like sewing," she says, and reminds me how when you're cutting out a sleeve, you have to allow for fullness around the arm. And I get it.

The next day when Art orders each of us to frame up our own dog house so we can cut our first rafters, I can hardly wait. But my rafter doesn't fit.

"You've got the right idea," Art says, "but you laid it out wrong on the framing square." Theory and practice. I see my mistake, and the second time I get it right. Both my rafters kiss at the ridge. Beautiful.

Cutting rafters is one of the marks of a real carpenter, Art says. Winders come the next morning, when he points out how my crew's floor plan didn't leave enough room for stairs to the second floor.

"Carpenters run into problems like this all the time. What shall we do?"

I love how he says "we." When no one else volunteers, I put up my hand. I'd read about winders the night before, and Art nods at me to explain how they're a kind of stair, pie-shaped and built in a spiral so they'll fit into a smaller space. Since I'm the one who knew about them, Art tells me I can build them. We've already built regular stairs, and by lunchtime I have two straight steps ending on a platform that will hold the winders. But I can't figure out the pie shape. Art comes to where I'm squatting on my platform.

"Remember, you're turning three steps in ninety degrees," he says, "which means each step will be what?"

"Thirty degrees."

He moves on to help someone else while I stare at the flat metal arms of my framing square, running my fingers over the numbers and tables imprinted there—impenetrable. A guy would say, "Look busy," so I pull out my hammer, drive some nails deeper, sit down again. I will not cry.

Art's back. "It might be helpful to draw where each of the steps will be," he says, hunkering down beside me. Immediately, someone calls him away. It goes like this all that afternoon and all the next morning. I crawl toward a solution. At noon on the second day, as the others leave for lunch, I stay behind, huddled on the steps of my platform, crying as quietly as I can. I can't do this. I'll never be a carpenter. When I hear footsteps I wipe my eyes, turn my face to the wall and inspect the plywood. After a few seconds I hear Art's voice. Damn.

"Checking it over before lunch, I see."

When I nod, he says, "I like the care you're taking. That's what school is for." To my dismay, he squats down beside me. "If you make your mistakes now, you won't make them on the job."

I can still only nod.

"I remember when I was in carpentry school," he goes on after a few minutes. "I had a hard time."

"You?" I say, disbelieving. Art has run construction jobs all over the world. Listening to his stories about the places he's built is my favourite part of every day. But now he tells me he did so badly in his first month at carpentry school that they wanted to kick him out. When I turn around to face him, he's staring straight ahead. He liked carpentry so much, he says, he begged his dad to let him try again, and the school gave him a one-month trial.

"What you like doesn't always come easy. Some people are born carpenters and some are made carpenters. I'm one of the made ones." He stands up and brushes sawdust off his pants. "They say it takes ten years to make a carpenter and it took me at least that, but sometimes I think it makes me a better teacher." Then he leaves, telling me to get myself some lunch.

Ten years. It takes ten years even when you love it. I have nine years and eight months to go.

—

On the last day of pre-apprentice class, as Jude and I are leaving the shop, Mr. Stewart waves for us to come back. Bob, grinning broadly, gestures toward one of the men's washrooms, where a small black and white sign on the door now reads "Ladies."

"I just went and bought it, specially!"

It isn't "Women," but it will do. One small step for womankind.

8

slippery, sloping surface

On Monday morning I head to the Manpower office, as Art had instructed us, and tell the counsellor I want a job as an apprentice carpenter. I have seven dollars in the bank. The man fiddles with some papers, then asks, "How fast can you type?" He isn't joking. When I say again I want a job as an apprentice, he says, "But you can type, can't you?"

I go to a different Manpower office and, this time, ask for a woman counsellor. She gives me a list of companies looking for carpenter's helpers or apprentices. Art said we should look ready for work, so I'm wearing steel-toed boots and carrying a tool belt when I knock on the first foreman's door. While I'm telling him I have two years' experience as a construction labourer plus a four-month pre-apprentice course, he's walking slowly around me, inspecting. I feel like an uncomfortable Venus de Milo.

Finally, "What does your husband think of you doing this?"

"I'm not married."

"What does your father think?"

At the second place, the foreman has some office work to do and asks if I can type.

The third place—a framing company called B&D—is more promising. The foreman, Brad, says simply, "We don't need anyone today," and my counsellor says to go back to the site every morning, since labour needs change daily. Soon it's routine: I turn up at B&D every day at 7:30 a.m. and look for Brad. Often I have to search for him, and once in a while I pass a man who's cleaner than most of the carpenters, wearing a white hard hat.

"A white hat," Art had said, "means an engineer or site superintendent." After a while, White Hat and I recognize each other and nod.

One afternoon, Dominic from the pre-apprentice class phones, elated.

"I just got a job with some framers! Great foreman named Brad."

It's the same Brad who'd said no to me again that morning. Ten minutes later I'm on my way to the B&D site, carried by fury. I have no idea what I'll do when I get there but before I find Brad, White Hat spots me.

"Come to start work?" he asks pleasantly.

"No."

He frowns. "I thought they hired someone today."

"They did. It was a friend of mine. A guy."

The super looks at me for a long minute. "I see," he says. "I have to walk that way anyway. You don't mind if I walk with you?"

We spot Brad in front of a half-finished house, waving a set of plans. He's surprised to see me, but before I can say anything, the superintendent smiles and says, "I hear you're hiring, Brad."

Brad looks suspicious. "Hired a guy this afternoon." He turns to me. "Too bad you didn't come a little earlier."

"Any chance you could use someone else?" the superintendent asks, still smiling. After a moment, Brad points to a man balanced atop a two-storey building nearby, directing a crane. I recognize the bony angles of trusses jutting over the edge.

"Think you could work up there?"

"Yes."

"She can start, then," he says, not to me but to the superintendent, "tomorrow at seven. Ten dollars an hour. Unless you don't want to work six days a week?"

"That's fine," I say, but I'm so excited I can hardly talk. I have a job! Putting roofs on. The one job I always avoided at school.

"Tomorrow, then," and Brad turns on his heel. He doesn't look happy.

The next morning, as I lace up my boots and drive to work at 6:30, terror is buried under the small voice that asks why they didn't hire me sooner, and how am I going to walk those top plates? Like a razor's nick, it bleeds all the way to the job.

The site is a development of about a hundred units, each with a concrete basement and two wood-framed storeys on top. As I park, the two houses in front of me bristle with labourers carrying two-by-fours onto a new platform, ready for framing. But something's wrong. Brad said they started at 7:00 and my watch says 6:45.

"You're late," he snaps behind me. "The boss likes you here early." He points down the road to the house he showed me the day before. This morning, the man on top shares the heights with a bundle of trusses.

"Dale," Brad says, and is gone.

Dale wears blue jeans, a white T-shirt and no hard hat. He's busy walking backward on the top plates, laying out where the trusses will go. In school, Art always said a good carpenter is ready for anything. He said some carpenters never get to build a roof in their whole lives. Right now, Art would say I'm lucky. When I reach the house, I'm shaking slightly with the chill of the morning. When the man on top doesn't notice me, I call up, "Dale?" My voice sounds small. No answer.

When I was eight, we moved into a new house in a neighbourhood that was still under construction. After supper, I used to love to slip past the night watchmen to explore half-built houses, sometimes with neighbour boys, more often alone. I'd pick out the room I wanted for mine, play with the booming of plywood under my feet and collect the silver coins under electrical boxes. Now I'm back, this time with a right to be here.

I clamber through what will one day be a front door as the sounds of Dale working overhead ricochet inside bare walls. The second floor is littered with bits of lumber, nails and empty nail boxes. Ray, Ted, Art—all of my bosses have been fastidious about cleanliness. It keeps a job safe, they'd said.

My head is barely through the opening in the floor when Dale barks, "I need spikes!"

I find an empty coffee tin in the rubbish on the floor and fill it from the fifty-pound box in one corner, then cross the floor to where a decrepit ladder is propped against one wall. When I climb it and hand him the can, Dale throws the contents into his belt with one quick motion and flips the tin to the floor.

"Up here," he orders.

"Here" is the two-by-four that forms the top of the framed wall. Dale doesn't look at me. He's bent double, walking backward along the plates marking sixteen-inch centres.

"Hurry up!"

I can't do this! But I'm damned if I'm going to make them hire me, then turn around and say I can't do it. And there's the matter of the seven dollars. So I climb to the top of the ladder, place my boots on the wall, let go my hands and slowly straighten.

The air up here is thin, and a strange temperature that swings between hot and cold. I smell lumber sharp as smelling salts, and I can see for miles out to where early morning mist still hides the airport. A breeze washes my cheeks. I am a tightrope walker with no net.

"Grab a truss!"

And now I'm being asked to fly.

Arms spread, each foot excruciatingly placed, I walk slowly to the end of the wall, ease myself into a crouch and lay my hands on the pile of trusses. I'm concentrating so hard, I feel only my hard-beating heart. Dale waits, casually perched on the opposite wall. I know my walk screams "novice," but I don't care. If I'm going to die here, at least I've lasted this long.

"The top one goes to the far end of the building."

I look behind me, horrified. The end of the building is sixty feet away.

He's already pulling at the top truss, bouncing it violently to free it. The whole building rocks, and the sound echoes in the drum of the framed box beneath us. Afraid he's going to throw me off, I slide back and, with one hand, hold onto the wall. With the other hand, I poke at the truss to help free it on my side. Dale half lifts, half pushes it along the wall while I follow, reaching out when I can spare a hand from balancing to give it a token push. When he finally gets it to the end he yells, "Three spikes in each!"

My first nail bends and takes forever to pull out again, braced as I am on nothing. I've just started the second spike when I look up; Dale has already driven three nails and braced the truss with a one-by-four and is turning back for the next one. And that's how it goes all morning.

When the coffee whistle blows at eleven, my legs are trembling from the effort of keeping my balance. As my feet touch ground, I understand people who leave boats and kiss the land. I stomp a few times, testing, then pick up my pack and follow Dale to his truck. I should be ashamed of my slowness, I know, but I'm pumped, thrilled that I've survived.

The van's double side-doors are open and three men inside are eating sandwiches, listening to a radio talk show. No one looks at me. Dale slides into the driver's seat. Should I sit beside him? Then the other fore-man, Dave, Brad's brother and partner, pushes past me and climbs into the empty seat beside Dale.

"Surprise me, big guy," he says. "Say something intelligent!" And switches the radio to a comedy station.

Murmuring excuse me's, I climb into the back of the van, sit down and pull the thermos from my pack. None of these men is over twenty-five. I am thirty-three. Feeling like a British matron who's wandered into a strip club, I hold my tea with two hands, elbows tight to my sides.

After the break, my efforts to help Dale grow even more token. Fi-nally he snaps, "Stay where you are!" I'm profoundly relieved to wait in one place as he hauls each truss along the building so I can just nail my side when it gets there. But now it seems silly to balance precariously as I wait. I climb down and set the ladder up against the wall so that each time Dale brings a truss, I only need to move the ladder along and climb up again, secure, waiting to nail. I try to be silent, like Dale, but once it slips out.

"Trusses seem so puny."

He doesn't even look up. Okay, if that's the way he wants it, I can be a machine, too: I nail. Just before noon, my right eye starts to water from the dust. I stand for a minute, blinking.

"Something in your eye?" It's the first thing he's said since coffee break.

"My contact lens."

But he's already fetching another truss, and I keep working, keep blinking. That's how I miss Brad.

"What's the holdup here?" Brad shouts as he comes up the ladder from the first floor. He does everything loudly. "What's taking so long?"

Any delay is clearly my fault. I look up at Dale.

"You know me, Brad. Thought we'd just take it easy," Dale calls down. Brad stares at me.

"What's this?" he snaps. "Why the hell aren't you up on the plates?"

"She lost her contact lens," Dale throws down, "so her balance is shot. She's half-blind."

I'm awestruck by the speed of his return, the ingenuity of it, his defence of me.

"Well, you better get yourself another contact lens awful fuckin' fast," Brad says, and disappears.

"Let's get this finished by supper," Dale says, and I nail harder.

By the two o'clock break, I can't wait any longer. When I ask, Dale points to a small green and white fibreglass closet. I've never used a Johnny-on-the-Spot, though I've seen them on construction sites. At the school, Ray had a flush toilet put in for us, and when I was working for Ted, I'd gone behind bushes, like the men.

I drop my tool belt and pull open the door. It's tidy. In front of me is a high white seat and beside it, on the left, a small sink. Then I notice the sink has no taps, is marked with yellow stains and holds a big white moth ball. Okay, this isn't a sink. On the back wall in black crayon is the rough drawing of a penis. As I latch the small hook that holds the door shut, the tiny space blooms with the stink of urine and disinfectant. Flies buzz. I sit with my feet dangling and wonder how I'm ever going to change a tampon in here. When I climb back into the van, the others are teasing the youngest helper, a big guy in a bright yellow T-shirt who holds his head and groans.

"One too many, Mitch?" someone asks, and they all laugh.

I recognize this from working on the island. Translation: he drank too much.

By three o'clock, the trusses have all been nailed and braced and Dale orders me to start passing up plywood for sheathing. I station myself beside the pile and pull the first sheet off, balance it on its four-foot edge on my steel-toe, then grab it at the bottom and push up. Dale catches it and nails down the first two sheets to give him something to stand on. Then we really get to work.

Once I have the knack, the sheets aren't that awkward; what's hard is pushing them up over my head. I'm discovering a whole new set of muscles across my chest and into my forearms. I know from my weight trainer that these are the muscles—pectorals—that give men the upper-body strength to heave sacks of cement over their shoulders, no problem. And most women's aren't as strong as most men's.

By the time we've done thirty or so sheets, I'm sweating, grunting loudly with each lift. I'm in survival mode, concentrating so hard I almost miss it when Dale says that's enough for now. I'd closed my eyes with the struggle of the last lift. When I open them, I catch the flicker of a smile before he turns away.

"Up here," he says, and I climb to the roof to help him nail.

Walking the slope of the roof doesn't bother me much except when I have to lean over the edge. When I stand up again suddenly, I wobble and almost fall. Dale, whose back is turned, yells, "Crouch! It gives you better balance." I wonder how he knew, and am grateful. He also mumbles a warning about the sawdust sprinkled around where we've cut the ply.

"Slippery as marbles," he says.

A few hours later, out of the clear blue, he says, "You know how you can tell a good carpenter?"

I'm overly eager to find out.

"When he falls off the roof, he yells 'Whee!' all the way down."

All afternoon I wait for him to talk again, but he only growls when I lay plywood the wrong way or put in too many nails. On the island I'd come to enjoy the men's non-stop one-liners, and here I miss it. This job is just different, I tell myself, and concentrate instead on staying upright on this slippery, sloping surface.

⌐

I come home that night, and every night after, exhausted, but there's something clean about framing. I work to my maximum physical capacity, and at the end of each day there's a roof, a wall, a floor—something I can see, solid and undeniable. I've kept notes every day since I started but in my need to get to bed and sleep, I make them less often now.

Fatigue is bad for relationships, too. "No sex life from Monday to Saturday on this job," a carpenter says one day at lunch. "Too tired." It's the most personal disclosure I've ever heard on a construction job, but at least I'm not the only one.

After a few weeks with Dale, I'm sent to work with the main crew. When I go for my level after we've lifted the first wall, Brad snaps, "What do you think you're doing?"

I know from school that framed walls must be perfectly plumb so that trusses, drywall, doors—everything that comes after—will fit properly. "Square and plumb!" Art drilled into us. "Square and plumb!"

"Get that out of here!"

I stand back as Brad gives the wall a quick eyeball. "Good enough," he calls. "Nail it!"

"But..." I shut up as he jerks his arm as if to hit me with his hammer.

"Do something useful."

I don't say anything back, not now, not any of the times when the men get impatient with me, ignore me, yell at me. After all, I tell myself, it's their territory. Instead, I bring my anger and confusion and frustration home and heap it on John and Kevin. Mostly John. And every time I come home complaining—again—about someone not treating me right, John says, "The guy's just a jerk. Don't take it personally. Tell him to fuck off." But I don't. I can't.

There's one thing I can do, though. I notice at work that when I call for studs, the labourers often don't hear me, so one day when I yell, I lower my voice. "More studs!" I call in my new bass, and one of the labourers yells back, "Will I do?"

It's in this rough way that the men also flirt with each other. One day Dale is in deep conversation with Ken, the new and drop-dead-beautiful helper, but as Brad walks by, Ken breaks off to say, "Dale's making sexual come-ons to me, Brad, but I'm resisting."

Brad's tone when he replies is casual. "You're a fuckin' cunt, Dale, comin' on to young boys."

After I get over my shock, it confuses me. Someone tells me they hired a woman here once but fired her because the wives were jealous; some days I'd say what the wives should be jealous about is not other women on the job, but other men.

This work should be straightforward: I turn up on time, I carry, I lift, I nail, I go home. But it's as if there's a vacuum around me; even when I go into the grocery store on my way home to pick up dinner, I can feel eyes disapproving of my dirty jeans, my man's shirt, my workboots. When I look, no one's watching.

—

My salvation is Women in Trades. We've met every month since that first meeting a year ago. Most of us are over thirty and have worked at our share of the usual "women's" jobs—clerical, service, child care. Several have university degrees and most of us have no kids, though we never talk about why. We send letters to companies and government ministries saying we're available if they need ideas for getting more women into trades, and some of us are invited to sit on committees and give speeches.

Mostly, we tell each other stories, like the single mother on welfare, sick of being poor, who tells us how—when she heard about the good money tradesmen can make—marched into Manpower and asked for trades training. Her counsellor was horrified when she said she wanted whichever trade paid the most.

"You can't pick a trade like that!" he told her. "You have to have a feeling for it."

So she went home, did her homework and came back, making sure to get a different counsellor.

"I told him that my father was an electrician, my brother was an electrician and I'd always dreamed of being an electrician. I practically said God told me to be an electrician!" She is now an electrician.

When some San Francisco tradeswomen start a magazine called Tradeswomen, several of us get subscriptions. The pictures alone give me hope: black-and-white photos of women ironworkers, carpenters, boilermakers, plumbers.

⚊

Every day I pack the biggest lunch on the crew—two sandwiches on brown bread, two pieces of fruit, cookies, a thermos of tea—and still I grow thinner. We're working sixty-plus hours a week at continuous bending, lifting, carrying, pushing, pulling and hammering. My arms and belly and thighs feel firm and tight. And it's getting better. Now that the crew are used to me, the talk that bubbles around us helps pass the day. When I try to join in, my comments sound stiff and self-conscious, so mostly I listen. I'm scrupulous, too, about hiding all signs of having a period, carrying a baggie to take my tampons home so they can't float to the top of the tank, as if the men are a pack of dogs who mustn't get the scent of blood. I expect nothing from them, am grateful for everything, feel safest when they ignore me and talk about things that don't interest me, like sports and hot cars. But slowly, they give me a place. Once, when a new guy sits in my usual space in the van, Mitch tells him, "That's Kate's spot."

Maybe that's what gives me the courage to buy a brilliant orange T-shirt with a roaring tiger's head embroidered on the front, at the Sally Ann. On impulse, I also buy red henna hair dye and my head comes out of the sink matching the orange T-shirt. When I get to work the next day nobody says a thing, but all day Dale calls me "Tiger."

Brad's youngest brother, Reg, turns up late one Wednesday morning. When he sees me he comes so close, our boots almost touch. He smirks, then, eyes glued to my face, says, "Girl on the crew, eh? This should be interesting!"

I've come to accept being invisible, and now I yearn for it as Reg seeks me out. If materials aren't coming fast enough or a nailing pattern is wrong, he looks for me to blame. One day he lets loose a string of obscenities, then turns and says, in an exaggerated tone,

"Oops, pardon me! Lady present!" Then, "But of course you couldn't be a lady after hearing all this."

"Nail every four inches!" he orders one afternoon. An hour later, as I crawl on my hands and knees, carefully pounding a nail every four inches into the aluminum strip between units, he snaps, "Why so many nails?"

I cringe. "Because you told me to."

"Oh, yeah. I just wanted to be sure you knew what you were doing." His lips bend into a thin smile, like a wolf's. "Hurry up."

When Reg is around I make mistakes, like the time he and Brad are watching me cut sheathing. There's an abrupt halt to the screech of my saw, and when I press the power button, nothing happens.

"You cut the cord." Brad's tight-lipped. "Get another saw." Reg just smiles the wolf smile.

I don't think I make more mistakes than the men, but if I even hesitate, Reg is onto me. "Are you an asset or a liability? Any more mistakes, and we'll have to let you go."

I come home every night seething. I only realize I've been worrying out loud when one night Kevin greets me with a cheery, "So what mistakes did you make today?"

It's how they get along with each other that feels so strange. Like the time Mitch nails a wall together wrong, and Brad explodes.

"Jesus Fucking H. Christ! Can't you read?"

Mitch corrects it, bashing savagely at the lumber, but after work, he and Brad head off together for a beer, best buddies, and I'm amazed. John, my primary if reluctant source of information on men, is no help. "They don't take it personally," he says, for the hundredth time.

But I am zombie-like with fatigue and frustration and fury, and now I lose it. I go hysterical, screaming and tearing at my hair, turning in useless circles in the living room. "How can I not take it personally if someone yells at me? It makes no sense!"

Finally, finally, he gives me strategies. "Don't wait," he advises. "Come right back at them the minute you don't like what they've said. Tell them to fuck off—say it loud. Avoid the worst ones. Bash your hammer around a lot when you don't like what they're saying. Talk to the supervisor who got you in there."

And yet I love the work. On the good days, I have the pleasure of working physically with these men to make houses appear, like on the June Monday about four months into the job when Mitch and I are working with Brad. It's a clear day with just enough wind to keep a sweat off—a perfect day for framing. I help Brad lay out plates so Mitch and I can bring studs, six at a time, laying them out according to the code scribbled in Brad's red crayon. Then we nail, two spikes per stud into each plate in a rolling motion of fingers, nails, arms and body until the wall is a solid thing under our hands. It takes the three of us to raise it, and suddenly there's a window to frame the Vancouver skyline. For the next few hours we work fast to see how many rooms we can build before lunch.

With Reg nowhere around, I relax a little, listen in on their banter, shelter the fragile feeling that I'm one of them.

"Shot a hole-in-one at golf last Sunday, Brad."

"You expect me to believe that?"

"Would I lie? This is on the level."

"Only time you're on the level, Mitch, is when you're lying down, and then you're probably humping something!"

Mitch glances at me. "Can't you keep your filthy mouth shut when there's a lady present?"

But I don't mind the talk. What makes me uncomfortable is Mitch feeling he has to protect me. We're in the middle of raising a heavy exterior wall when Reg suddenly appears and notices we forgot to cut short the top plate. With the wall half up, he yells for me to grab the saw and cut the plate—now! But it's at an awkward height above my waist, and shaky. With Reg, Brad and Mitch struggling to balance the wall in midair, I pull the trigger on the saw, then realize that at this angle, the guard is in the way of the cut. It's the only time I've ever broken the "First, look for the fingers on your left hand" rule that Art drilled into us at school. To go faster, I keep the power button pressed down, blade spinning, while I reach blindly over with my left hand to raise the guard. The tone of the saw suddenly changes, and I feel something warm on my fingers. I know that when I pull back my hand, it will be covered with blood.

"Someone grab that saw," Reg yells. Blood streams from my little finger. The end of it hangs at an odd angle.

"Are you okay?" a voice asks.

I pull out the blue cotton handkerchief I keep in my pocket to hold back my hair and wrap it around the finger. "I need to go to the hospital."

They're busy raising the wall. I unwrap the handkerchief and inspect the bleeding lump as if it belongs to someone else. It doesn't hurt, but the end of my finger dangles by a small red thread.

"I need to go to the hospital," I say again, then turn and walk across the deck toward the ladder that leads to the street. My legs feel distant beneath me. I see Dale, and suddenly I want to cry.

"Go to the First Aid shack," he says.

But the First Aid shack is empty. After what feels like a long time, one of the labourers rushes in, but there's no tap, no water, and as he searches for the peroxide, I tell him not to bother with a bandage. Slowly, as if a fog is lifting, I'm getting mad.

Dale is back, offering to drive me to Emergency, but I'm mad at him, too.

"I'll drive myself!" Suddenly I can see how crazy the last few weeks have been. We all pretended nothing was wrong, but now I can point to a piece of dangling flesh and say, you see? It wasn't all in my head. What Reg was doing—riding me—was dangerous.

The doctor is more interested in my steel-toed boots than in my finger.

"Is this what you wear to do your housework?" he asks, chuckling at his own joke.

This is what I wear to kick smartasses around, I want to say, but don't. I'm seething at Reg, at Dale, at men in general, doctors included.

"A flesh wound," he finally reports. "The bone is okay. You're lucky." He removes the fingernail, stitches me up and tells me to take the rest of the week off.

When John sees me on the couch with my left hand swathed in a bandage, he's furious. For weeks he's been begging me to quit this job, but I wouldn't listen. How could I? Reg would think he'd won. So we have another in a long line of arguments that ends with me yelling, "I love this work!"

John points at the bandage. "You love that?"

He's afraid for me, I know, feels helpless, but I feel like he's saying I can't look after myself. We eat another silent supper.

On Friday, when Brad phones to tell me to pick up my paycheque, I time my visit for the morning break so I can show off my finger, looking like something from outer space without the nail. I feel tough; I'm blooded now. Brad's friendly, walks me back to the truck.

"We're wrapping this job up," he says, though I've just seen two more units started. "So we won't be needing you anymore."

I'm laid off. I'm calm, like a cow placidly watching the sledgehammer aimed at her forehead.

9

shameless

My counsellor at Manpower gives me a list of companies who are looking for an apprentice, and I spend the afternoon on the phone.

"Sorry," they all say when they hear my voice, so I look up "Construction: Contractor" in the Yellow Pages and start at A. When I get to the Cs, Jac Carpay clearly doesn't want any more to do with me than the others. Still, he doesn't hang up right away, and even agrees with me that a woman has to work harder. His daughter is having a heck of a time, he admits, breaking into management at a local company.

I make a sideways move. "What kind of building do you do?"

He counter-moves, trying to end the conversation. "Renovations and additions, but nothing's the same two days in a row."

Finally, after I offer to work for a day for free, he says I can come and look at the job. "Then, if you're still interested," he finishes, "we can talk."

If you're still interested, I think, and hang up.

By now my truck has utterly died, so on Sunday night John drives me over to drop off my tools. But it isn't until the next morning when I get off the bus that I can see the unfinished slash of wood against blue sky. The house under renovation looks as if Godzilla has bitten off its right-hand side. On the left, the neat dark brown of the original stands like a doll's house, half the rooms open to the air. I walk up the sloping board that takes the place of stairs, onto a plywood floor.

"Hello?"

My voice squeaks. I hear boots as a man comes around the corner.

He's scowling.

"I'm Jac," he says, and sticks out his hand. I like him instantly, even if he is scowling. He's short, and older than I expected—about fifty, clean-shaven with a head of thick brown hair. He's solid in build, with quick, sharp movements.

"You better have a look."

As he guides me through the rubble, he explains that the man he's been training as an apprentice has just moved out of town. The family who own this wreckage are managing to live in the still-standing half, and Jac introduces me to the woman, who's sweeping sawdust from her half-demolished kitchen—a job about as hopeful as sweeping sand off a beach. At the front of the house, as I turn to shake his hand goodbye, Jac says, "If you're willing to do your share of the work and learn the trade, then I'm willing to pay a fair wage and teach you for a three-month trial period. We start eight o'clock Monday morning."

It isn't until I'm halfway home that I remember we haven't discussed money.

—

Monday morning I'm up at six so I can get to the job—a long bus ride across town—by eight. I'm getting used to people staring at my steel-toed boots and grey worker socks with the red stripe, carefully pulled up over my jeans to keep them from catching. This morning I stare back, proud. And nervous.

Jac is already there, giving instructions before I've finished buckling on my tool belt.

"I'm going to the lumber yard for supplies. While I'm gone, I want you to take this door out and frame in a new one. Measurements are on the plans."

I've never put a door in. Still, what could be easier than pulling something apart? I collect the tools I might need. This will have something to do with force so I hoist my hammer and pick up the crow bar—no, too big—pry bar, then, and set it to the outside trim. A few taps, and the edge lifts. Feeling oh-so-competent, I can now see the bones of the thing, the two-by-fours that hold the door up on each side. I take my nail puller to the door jamb, hesitating as I gouge into the finished wood—did he say he wanted to save it? But the nails are old, with no head to speak of, and they're stuck like King Arthur's sword in the rock.

I crouch close, one leg on either side of the door, centre my weight in my lower body and heave. When there are only three nails left, a strange thing happens; the door tilts alarmingly, just as Jac returns. He surveys the scene—wreckage, really—and scowls.

"Might be easier if you took the door off its hinges first."

"Good idea," I say, a little breathless. I pull the pins from both hinges, and the rest is a piece of cake.

After lunch, Jac asks if I know how to lay out a wall. Much of the ease of building depends on layout being accurately—and quickly—done. At B&D, it was saved for the foreman. But I studied it in school, even laid out one wall, marking where doors and windows and studs should go.

"Sure," I say.

Later, as I demolish the rest of the kitchen, I watch out of the corner of my eye as Jac checks what I've done. He calls me over to show how I've laid out the first stud wrong, not allowing for the ply having to run three-quarters of an inch long, to cover it. My mistake would have thrown off all the centres when we came to sheath and drywall. But Jac acts as if it's no big deal, so I ask a favour. It takes nerve, but this is important.

"Next time you ask me to do something new, could you show me how? If you show me once, I'll know how to do it right."

By the look on his face, I'm certain no man has ever made such a request; a man would fake it. But Jac's in it this far, and he goes along. It gets to be our joke. Every time something is harder than I think it should be, I yell, "I need a trick of the trade!" and Jac stops whatever he's doing to teach me another of the shortcuts that make a job easy: testing for stud location where your finish board can hide the nail holes; marking the measurements for a new sill by laying it on a temporary shelf in front of the window opening; and never, ever assuming anything is square or level.

He's Dutch, trained in the European system of seven years as a journeyman carpenter instead of our four. Maybe that's why he takes my training so seriously. Every time there's something new, it's "Kate, get over here!" and I know I'm about to learn something good: how to take out a window, how to fit a shelf, how to hang a door. One day I notice I don't approach new things with trepidation anymore—I look forward to them.

"Kate, get over here!" And I go, eager.

One night when I have to phone him at home to ask about hours, his wife answers. She's cool, and I'm careful to tell her who I am and why

I'm calling. I'm not after her husband, I just want to work for him.

It's typical that when Jac makes the big announcement three months later, he's casual.

"You better get those papers," he says.

What papers?

"Apprenticeship."

A few days later, John and I sit formally in Jac's living room as he spreads a sheaf of official-looking forms over the coffee table at our knees. Under the provincial system, an apprentice's rate of pay is set by the journeyman's. A first-year apprentice starts at 50 percent of journeyman rate, and every six months gets an automatic raise. After four years, and six weeks of school every year, the apprentice makes journeyman's pay. I'm considered a second-year apprentice because of my labouring experience and pre-app course, and Jac has set the journeyman rate at ten dollars an hour, so my starting rate will be six. Every other apprentice I know at my level is making more, but I let it pass. I love working with him and he takes my training seriously; that's worth a lot.

Besides, I'm getting the feeling that Jac looks out for me. One day, when two roofers show up to give him an estimate, they elbow each other and leer in my direction. At coffee, I keep my voice casual as I ask Jac if they're the ones who'll get the job, and when he says no, I dare to ask him why.

"I didn't like the way they acted around you."

I do seem to make men nervous. Even after four months, the man who owns this house—Fred—watches me like a hawk. I learn one useful thing from Fred. Any big obstacle on a job demands silence and concentration, but I've got into the habit of swearing, a mild "Shit!" every time I meet a small glitch. It's fun, like swaggering. One day Fred overhears me.

"Is there a problem?"

By the third or fourth time he asks, I get it: my swearing makes him nervous. It's his house, his money, and I'm the first woman he's ever seen do this work. So I stop swearing.

Years later, when I read a book by Norah Vincent, a woman who went "undercover" for eighteen months to pass as a man, I'll think back to this job. Vincent didn't just have to cut her hair or fake a five o'clock shadow; she had to learn the body language of men, including how not to stick her neck out. Literally. When women talk, her acting coach told her, we lean forward, straining our necks. It's one reason our voices are higher. Vincent had to learn to control her gestures, use fewer words, keep her voice low, not touch people. And that's how it is with Jac. He's teaching me how to act like a carpenter: don't smile too much; don't talk, or at least not in consecutive sentences, and then only to joke; never touch anybody unless it's a slap on the shoulder. If I forget and laugh out loud, Jac will say, "Now you're acting like a woman!" only half-teasing.

Not every day is a good day. On days when I can't measure right, don't know what comes next and generally feel useless, I wonder again if I'm too slow or just too stupid for this job. Are women really less "natural" at this than men? It helps to remember the last Women in Trades meeting, when Judy confided that if anything went wrong on a job— something as simple as a nail not going in—she knew it wasn't because the wood was old and dry, or because there might be a knot there, it had to be because she was a woman. We'd laughed because it was so ridiculous. And because we'd all felt the same.

One Monday, when Jac gets back from picking up materials, he's scowling—always a sign Jac's about to tell me something important. He takes my old hammer out of my hand and replaces it with a beautiful bright red one.

"That one of yours," he says, "is a piece of garbage." The new hammer weighs twenty ounces instead of the sixteen I'm used to. "It's a Plumb," Jac says, and shows me how to measure its balance by setting it on its head and, most thrillingly, how to get the feel of it with a few practice swings, like the men I've seen buying hammers in hardware stores.

"Nice colour," I say, and he checks to make sure I'm (sort of) joking.

"Now I expect you to nail twice as fast," he says, as if it's his turn to joke. We both know he's not. And sure enough, that afternoon he calls me into the living room.

"Let's see what you can do with that fancy new hammer."

With the roof on, we're ready for what he calls back-framing—filling in bits of two-by-four so the drywaller always has something to nail to. It's fidgety work that I don't like nearly as much as the big-boned bossiness of framing, but it has to be done, and it's the apprentice who does it. Jac points up to the ceiling joists and tells me to chalk a line every seven feet and nail two-by-tens into the spaces.

"Blocking," he calls it when I ask. "To keep the joists stabilized."

"Torture," I think as I balance on the tallest ladder, hammering over my head. Jac has shown me how to nail this way, forearm parallel to the floor, thumb below the hammer, but these are three-and-a-quarter-inch spikes. Lots of them. No matter how I hold the hammer, working against gravity like this hurts. It's those pectorals again. My progress slows to tap-tap-rest. Tap-tap-rest. Until I hear a squishy sound I can't place, and when I look down, Jac is standing at the living room door watching me, bursting with trying to keep his giggles quiet.

"You have a sadistic sense of humour, Jac Carpay!"

To which he replies with a full belly laugh.

"Keep going. You're doing fine!" He retreats to the kitchen, snickering.

—

In carpentry, I'm learning, there are five ways to approach any job, ten ways to do it right, and a hundred ways to do it wrong. I make decisions every minute, starting with something as simple as "Should I use two two-and-a-quarter-inch nails, or one three-inch nail? Angle them, or drive them straight?" Every mistake makes me nervous, even though Jac repeats many times, like Ted before him, that every carpenter makes mistakes; it's how you fix them that matters.

No matter how impossible a job seems at first, I'm beginning to have faith that, as Jac says, there's always a way. He calls it "the beauty of mechanics" and says that if something doesn't work, there's always a reason. This is a different kind of faith, the kind that makes me feel immediately stronger: faith in the materials, faith in the process, faith in the mechanics of building. Still I haven't yet got the pace.

"Slow down!" Jac told me often in the first few days. "You're going to hurt yourself!" But lately he asks—no, pleads—for me to "Please try and go faster! It's not a piano!" When I apologize for being slow, he says, "Don't worry about it. First learn to do it right, then you can get fast."

"Apprenticeship," Heather reminds us at one of our Women in Trades meetings, "is a ticket to know nothing." But I want every job to be tidy, neat, perfect. That's the nice girl in me, colouring carefully between the lines.

After Jac and I finish insulating and putting up the vapour barrier—a plastic sheet that keeps air from passing through the wall—an odd thing happens: the inspector is scarcely out the door before Jac hands me his utility knife and orders me to slash the plastic. I look at him, astonished. Even I know that the whole point of a vapour barrier is that it be airtight.

"Especially the kitchen and bathroom walls," he orders, "where humidity will be highest."

Obediently, I put neat slashes across the walls. "Twenty years from now," Jac boasts grimly, "our clients won't even know how grateful they should be, while their neighbours get mould and rot from a building too well sealed."

—

Women in Trades groups have been forming across the country, and one Friday evening in the fall of 1980, eighty tradeswomen gather in a hall in Winnipeg for the first-ever Women in Trades conference in Canada. As women's voices call out their trades, the atmosphere is electric. Carpenter! Electrician! Plumber! Ironworker! Sheet metal worker! Boilermaker! Welder! Electronics technician! Machinist! Even the entertainer, Heather Bishop, is a tradeswoman. When she sings about a female machinist's reaction to a man who touches her behind—"Buddy, what you're lookin' at now is a woman's anger," we cheer and hoot, and by the end of her performance, we're all standing in ovation to her and to each other.

There are workshops all day Saturday, and on Sunday morning, a fashion show. It sounds like a weird idea, and when the first woman hits the runway dressed in railway coveralls, there's a moment of silence— what is this? Then the commentary begins. "Notice the slimming effect of the thin vertical stripe, the concealing loose coverall, a dapper touch of red in the neck scarf." The room erupts with laughter and applause. That's my life up there, that woman in the tool belt swinging down the runway, proud.

Until now, I've never had much contact—that I was aware of—with lesbians, but on Friday night after the entertainment, a ferry worker invites me into their circle. We find the club we're looking for in a dark back alley—there's no neon, no sign over the door. The name of this place is passed around only by word of mouth. When we knock, a hefty-looking woman opens the door and looks us over before we're allowed in.

At first, the crowd inside looks like any other. Some women are heavily built with short hair, plaid shirts and blue jeans; others are petite and conventionally beautiful, dressed in silk and wearing makeup. But a woman tends bar—it's the first time I've seen that—and there are women, only women, at every table, only women on the dance floor. When someone asks me to dance, I'm nervous. It's okay as long as it's fast music and only our hands touch, but when a slow piece comes on, I feel the urge to bolt. Too late—she's already got one hand around my waist. I fumble a bit, then put my left hand on her shoulder, my right hand in her hand.

It's different from dancing with men. She's shorter than most men, for one thing, and when I feel the occasional pressure of her breasts against mine, it's as if the energy off her body is less different, less distant. When I look around, which I do to the point of rudeness, I try to pin it down. These women don't curve inward as if protecting their merely female heart, female breasts. Their shoulders are back. They stand straighter, the lesbians, than so-called straight women. Still, I'm relieved when we all head back to the hotel and the conference.

There's another difference that becomes clear over the next two days as I hang around with the lesbian women; in two days, we barely mention men. There's no talk of male partners, no complaining about male behaviour, no talk even of the men at work. In this world, men don't matter. I see in the gap how much I've given my full attention to men, how I've lived for their opinions even if it's been in opposing them. The lesbians tease and appreciate each other—and me—as if there were no shame in being female. They even half-jokingly invite me to join a few of them in bed. I say, a little too politely, "No, thank you," and they laugh.

It will take me a while, but eventually I get it. No shame. Kate Millett talked about it, too. Before this weekend, I had no idea how deeply shame for my woman's body had been ingrained in me. Each message in itself was small; it was Eve's shame, or the shame of not being thinner, or prettier, and "down there" (we barely dared name it) was "dirty" and not to be talked about except for medical purposes. As a feminist I've been angry for years about women getting paid less than what men get for the same work, about women not being CEOs and carpenters, but this underlying shame about my body is more subtle; it's as if a small annoying bell has been ringing in the background all my life and I haven't noticed until now, at the tradeswomen's conference, when it stops.

Its silence is alarming. This weekend my body feels as if it's coming alive; not exactly "sexy," though I feel that too, and am afraid. A new channel has been opened. What if it becomes a flood? Does this mean I'm a lesbian? It seems that for most of my life I've been trying to cut off my energy at its physical source, and now, being with lesbians, it has permission to flow. Deeper than sexy, this new sense of my body feels erotic, like an electrical source is erotic, like a Georgia O'Keeffe painting is erotic—more alive. I've always felt good in my private woman's body when I was in bed, having sex. And over the past few years I've realized my body can feel good doing physical labour, building. Now I know it's okay to feel good in my public woman's body, too. Does this seem like a small thing? To me, it feels revolutionary.

At the airport in Vancouver, John comes to meet me. One weekend away, that's all, and yet I'm sure he must feel it on my skin, that I'm a different woman now, a woman awake, without apology, without shame.

—

"Kate!" Jac calls from outside. Purring in front of the house is the biggest truckload of what looks a little like plywood that I've ever seen. Jac has that funny look I've come to recognize as trouble. "Kate, meet drywall!" He hasn't been able to find anyone else to do the job, he says, so, "You just became a drywaller!"

It's not so bad—sort of like putting together a puzzle, though heavier. Years later, when someone asks me for a time of epiphany in my life, I'll remember that I'd actually met drywall when I was about eight. I'd gone with my parents to inspect our new house, still under construction, and wandered off to explore. In a hallway I'd found a long, unbroken stretch of white that glowed, luminous in the half-dark, calling for me to raise my two hands and place them square on its pure white space. When I pressed my cheek to it, it had a faint, sharp smell like the bleach in Mum's laundry. Later I would know it was newly plastered, by hand. I inhaled, intoxicated, pressed my whole body against it. It was the softest thing I'd ever felt, yet when I put my shoulder to it and heaved, I found huge pleasure in its resistance.

"What are you doing?" Dad asked behind me, and I'd pulled back, startled.

Today, pre-mixed plaster is pressed between two sheets of heavy paper so that the whole thing, called drywall, can be nailed to wood studs, and only has to be plastered and sanded at the joins. Sanding plaster is gross work, especially overhead, but I'm beginning to learn that if you don't like something in renovations, just wait. And sure enough, when the sanding is done and the painters move in, Jac hands me another address.

"Our next job."

The following Monday morning, standing in the backyard where the new addition will go, Jac explains that these clients don't want their yard messed up with machines. "So we dig!" And he hands me a shovel.

All day it rains a miserable grey drizzle, and the next day it rains even harder. By now I know Jac well enough to get away with complaining, which I do. I refuse to take pleasure in the rhythms of the work, in the clean contours of a steadily growing hole. Instead, I complain about the rain, I complain about the mud, I complain about the flat tire on the wheelbarrow and then I complain about my shovel. Eventually we work in silence, spelling each other off on rolling loads of dirt to the growing pile in the back alley. Rain catches in the brim of my hard hat, so every time I bend forward it rolls off in a shower over my eyes. Scheming for warm flesh, it slides over my collar and drizzles down my back. At 2:30, Jac says, "I'll take you for coffee." Jac knows the way to an apprentice's heart. When I get to the coffee shop, still fully dressed in rain gear, I stand back to hold the door for a man leaving, who says, "Thank you, girl!" He spits out "girl," and at first I think he's making a point of not saying "woman." Then I realize he thinks I'm a guy, with long hair.

The next morning, Jac asks casually if I can handle footings by myself while he goes for materials. Footings are the concrete base on which a house rests. I've helped others, but never laid out or built one by myself. I try and sound more confident than I feel when I say "Yes." Of course. It takes most of the morning, and when Jac sees the results— a classic, built-to-the-book set of forms, a thing of beauty if I do say so myself—he says, "Good," and tells me what to do next.

This, too, I'm learning, is part of the code. You know you're doing well if no one's yelling at you, so for a boss to actually say "good" is high praise indeed.

On the day we finish, Jac takes me across the street for afternoon coffee break, so we can sit on a neighbour's lawn and admire the clean lines of the addition we've just built together.

"That's the house that Jac built," he says, teasing. "And Kate."

That night I come home in my grubby coveralls and workboots singing the hit musical song "I feel pretty! Oh so pretty!"

—

In the fall I get a phone call. Dad has gone to a treatment centre to stop drinking, and now he and Mum are coming across the country, visiting each of their six children. I'm curious when they invite me to their hotel room in Vancouver. Their manner is formal, as if they're about to announce something important, but all that happens is that, in a general rush of conversation that covers many topics, Dad mentions in passing, "My drinking didn't hurt you, did it?" I barely have time to register the question before we're on to other things.

Later, I find out that one of the twelve steps of Alcoholics Anonymous is to make amends to the people the alcoholic has harmed. All I can see on this visit is that my father seems ever so slightly more humble. Still, I'm suspicious.

—

In January 1981, while Jac is on holiday, I start my second year of apprenticeship training at school. My friend Jude has decided not to go on in carpentry and is doing teacher training instead. When I walk in, the rest of the class is male; I'm on my own.

One day at lunch, the guys get into comparing wages. Most are making ten to twelve dollars an hour while I make six. When I defend Jac and explain I'm getting excellent training, the others aren't impressed. I bite my tongue before I can add that he also makes sure I'm not hassled. They wouldn't understand. Still, when I watch them, they aren't doing twice as much work as me, or doing it twice as well.

A few days later, our teacher comes into the room, ashen, to tell us that four carpenters have just fallen from the thirty-seventh floor of a downtown high-rise called the Bentall Centre. We press him for information, but so far, he says, no one knows more than that. Over the next few days, we huddle in the cafeteria at coffee and lunch, sharing every morsel of news. There are rumours that with construction work getting scarce, workers are taking more chances, afraid to take the time to make a job safe. One of the four who fell was an apprentice.

The beauty of apprenticeship is that you're only in school six weeks or so (depending on the trade) and for the rest of the year you're working, earning money, putting theory into practice. I go back to work determined to ask for a raise, but Jac beats me to it. At coffee on the first day he says he's now paying me seven dollars an hour. I thank him, then tell him the other apprentices are making ten and twelve.

"Seven dollars is what I pay," he says stiffly, and tells me that's more than he was getting when he was an apprentice, though he was sixteen at the time and living at home. So a few weeks later, when I get a phone call from a man named Bill who says he's the Joint Board—joint employer and union—Apprenticeship Coordinator for the Carpenters' Union, I listen. In school, I'd applied to join the union as a routine thing, trying to keep all my options open, though at the time I'd pictured a filthy rich organization whose arrogant leadership ordered its members around. Now Bill tells me the union needs good apprentices, and he hits a nerve when he asks how much I'm paid.

"At union rates an apprentice at your level is entitled to fourteen dollars an hour, plus benefits," he says.

But there's more to this than money.

Union members do mostly commercial and industrial jobs, he continues, three-storey walkups and high-rise office buildings and bridges. Finally, he says the Vancouver local doesn't have many women apprentices.

"In fact, you'd be the first. We'd like to see that."

This is a change. My friends at Women in Trades think joining the union would be a good idea, so—feeling terribly disloyal to Jac—I agree to meet Bill at the union hall after work one day, "just to talk."

My grandfather on my father's side was an ironworker, but otherwise I come from a long line of farmers and no one in my family that I know of has ever belonged to a union. "Greedy unions" were among the things Dad and I used to fight about. Now I follow Bill past fake-wood-panelled walls to find Alex, a business agent (what is a business agent?) who's buried in a crowded cubbyhole behind piles of fake-wood-printed boxes. The things I've heard and seen on TV about unions sit on my tongue like lead. Certainly, their offices aren't as luxurious as I'd expected. After Alex brings coffee, we sit in awkward silence until he tells us in a rolling brogue how as a teenager on his first job in Scotland, he'd been terrified by an older woman who ribbed him and threatened to pull his clothes off. I begin to relax.

"Our motto is 'A fair day's work for a fair day's pay,'" Alex says. "The union isn't a monster. It's you, if you join, and me. Every member has a vote on everything, including the contract. All we ask from our employers is fair treatment."

When I ask how that would apply if some guy was giving me a hard time on the job, he responds with a term I've never heard: job steward. Every job elects one man—"er, member," he corrects himself—to be a steward, and the steward's duty is to speak up if there's a problem. "Any problem," Alex says, "including harassment." It's the first time I've ever heard that word, too.

For weeks I worry about what to do. I'm curious about what I'd learn, and I could sure use the money. I miss being on a crew, listening to the banter. And the idea of a union intrigues me. Rambunctious, spirited, not shirking from controversy—it sounds like family.

There seems no way to soften it. The next time Jac and I go out for coffee, I take a deep breath.

"I have something to tell you."

"Oh, oh."

I feel terrible when I give him the news, and so does he.

"I figured when I signed that contract with you, it was for the full four years," he says.

"So did I, but things have changed." I love renovations, I tell him, but it's been almost two years and I want to try bigger jobs, work on a crew again and—with a stab of guilt—I'll make better money.

I feel even worse when Jac says, "This happens all the time. Little guys like me train an apprentice to the point they're productive, then the big companies steal them away."

"I'm sorry," I tell him, and I am. But I won't change my mind.

Two weeks later, I stand outside Vancouver Local 452 of the Carpenters' Union, ready to sign up for my first union job.

10

union maid meets concrete

The dispatch hall is about the size of a living room. In one corner is a coffee pot with a sign: Help yourself. Behind a counter stands a man in his early sixties, partly bald, wearing a saggy beige Perry Como sweater. His face rolls over the bones of his skull like putty as he inspects me over the top of thick dark glasses.

I step up to the counter. "Are you Steve?" Bill had told me who to ask for.

"I am."

"I've come for a dispatch."

"So you think you're ready for work, eh?" Steve tells me to sign the list in front of him, then pulls some filing cards from a worn box. "Have you been accepted into the union?"

"Not yet." Because there's a shortage of apprentices, Bill had explained I could be dispatched right away and accepted into the union at the next meeting of the Vancouver local, when the members will vote on it. Steve nods and goes back to his cards.

"You know if the members don't approve you, you have to leave the job?"

I know. My heart is pounding.

"Here's a good one." He pulls a card from the box. "Formwork. Can start tomorrow. Done any formwork?"

I nod, and he raises his pen, but there's something I have to clarify first.

"Could you give me a job with a good job steward?"

Steve lays down the pen, hooks his two thumbs over his belt and rolls back on the balls of his feet. "Why are you worried about that?"

I'm worried about another job like B&D. All it takes is one jerk with no manners and a big chip on his shoulder. I keep my voice firm. "If I have any trouble on the job, I want to know there's someone behind me."

"Sister," Steve says, looking me in the eye, "in this union, if you have any trouble, we're all behind you."

Five minutes later, as I walk out of the union hall with a yellow dispatch slip in my hand, I hear him chuckle, "Here we go."

—

At home that night I bring my tool belt and tool box into the kitchen. With every paycheque I've bought a new tool, as Art suggested in school, and each one is marked with a dab of the reddest nail polish I could find. I remove everything and start to sharpen and oil, placing each tool back in the box or in my belt when I'm done.

Minding your tools, Jac has taught me, is basic maintenance. I run a finger along the razor-sharp teeth of the handsaw, then check my pencils. Carpenters' pencils are rectangular so they don't roll. A journeyman once told me he could tell a good carpenter by her pencils, so I sharpen mine with long, careful strokes until the points are clean. Work with me, I'm asking my tools. Don't leave me with a rough edge or a bleeding hand. Then I iron my favourite work shirt.

When I wake up the next day, my stomach is so knotted I'm afraid I won't make it to work. What if this crew are like B&D? I'm already sweating by 7:00 a.m. as I haul my tool box through the front gate, past the sign that says "Hard Hat Zone." The dirt and gravel in front of me are strewn with the debris of construction—short lengths of two-by-four, bits of reinforcing bar, abandoned cement bags. In the first-storey walls, black holes gape like eyeless sockets where windows will go.

"Look for the foreman's shack," Steve had said. The first trailer holds a big table surrounded by four men who freeze as I walk in. When I say "carpenter," one of them points to a second shack where a tall man in a white hard hat almost collides with me on his way out. I thrust the dispatch slip at his hand. How do people introduce themselves?

"I'm reporting for duty. Carpenters' union. Apprentice."

The man says, "Oh!" then weakly, "Good."

It isn't ribbons and balloons, but it isn't rejection, either. He calls over a foreman, who says, "Find a place in the shack for your gear," then tears off one part of the dispatch slip. "Give this to Dennis, the job steward." The blue label on his hard hat says "Simon." As I take the extended slip, he says, "Welcome to the job."

When I appear in the doorway of the carpenter's shack, conversation stops dead. Then someone says, "New guy, eh?"

"Hi," I say, and step inside.

One of the younger men points at an empty spike, where I hang my belt and rain gear. I add my tool box to the collection on the shelf at one end of the shack and sit down.

"You won't need all that here," the young one says. "Just your hand saw, level and framing square." When he's interrupted by a loud horn, I grab my hard hat and hand tools and follow him outside to where Simon's waiting to lead me through the dimness of the first floor of the building. I'm careful not to trip over the short metal posts of rebar that jut out of the concrete like rough teeth. Eventually these will become the joins for new walls, but for now, they're capped with lengths of two-by, in case anyone falls on them.

Simon's already halfway up a wooden ladder to the next storey.

"Jergen," he calls at the top, and a grey-haired man looks up. "Meet your new partner."

Jergen moves over to make space for me on the outside scaffolding. I climb over the half-built wall onto a plywood platform suspended over the street, but I'm barely on the other side when someone calls my name. It's the job steward, a young guy with dusty blond hair sticking out below his hard hat. As we shake hands, I'm grateful that even if the whole union is behind me, it has a human face.

"Any questions, come see me," and he leaves.

Jergen isn't much taller than I am. His body is round, like one of those Russian dolls that nest inside each other, and his motions are slow and deliberate.

"Did you ever work with steel walers?" he asks politely.

The noise and traffic of the street below are startlingly close, but the four-foot-wide plywood platform we're standing on (a highway, someone calls it) boasts not just a handrail but a mid- and a toe-rail of two-by-fours. I take out my hammer while Jergen explains what we're doing, though at first I hardly hear him for the roaring excitement in my ears. He's tremendously patient. Each time there's something new, he tells me what we'll be doing, shows me, and finally he lets me do it, watching to be sure I've done it right.

"Well done," he offers the first time, and I almost fall off the scaffold in surprise. Over the course of the morning, several men walking by make a point of saying hello, asking how I got into construction, how I like it. One shakes my hand and says, "You should be proud of yourself."

All morning I follow Jergen like a good apprentice, doing as he does, working flat out as I lay steel beams across metal snap ties, tapping them securely with wedges to the plywood walls. Several times Jergen looks at me oddly. Finally, he says, "You don't have to work so hard, you know."

I'm confused. Perhaps it's true, after all, that union workers are lazy time wasters?

"Twenty years ago we used to have to work like that," he says. "If you didn't, you were out. But it's not like that anymore. Now we think you should be able to go home and still have some energy left for your family."

For the rest of the day, every time someone stops to talk to me, I get nervous that I'm doing nothing. But no one yells, and I start taking more time, especially for safety. This place is more dangerous than anywhere else I've ever worked. Here, we're always working at heights, with heavy materials, while a crane swings overhead carrying plywood and steel. I try not to think of the apprentice who fell.

When we've almost finished the first section, Jergen asks our labourer to bring more materials. What a luxury, after all the years I've been doing the fetching myself. All I have to do here is wave my hand for more ply, more nails, more steel. Actually, Jergen does most of the waving; I just fill my pouch with nails and keep my hammer busy. Head down, ass in the air.

Everything on this job is big: there are coated three-quarter-inch sheets called form-ply—hundreds of them—and long steel snap ties with plastic cones, wood studs and steel walers to hold the forms against the weight of concrete, and a huge crane chirping high over our heads. When the coffee bell rings at ten o'clock, the noise on the site stops instantly, as if someone has waved a wand. I slip my hammer into its leather strap and fall in with the flash flood of plaid shirts that funnels down the ladder and into the lunch shack. I've barely begun my sandwich, tea steaming beside it, when the bell rings again, and just as suddenly, everyone rises and sweeps out of the shack like a flock of dark birds. At the door, one of the men has hung back a bit.

"It's nice to have a woman on the job," he says.

When I climb back up the ladder, Jergen's already at work.

"How long was that break?"

"Our contract says two ten-minute breaks and one unpaid thirty-minute lunch break," he says, "and that of course includes getting to and from the lunch shack."

On every other job I've worked, "ten minutes for coffee" was elastic. Here, we get what we bargained for. Literally. I remember my father's talk about "unions getting away with murder," but obviously it wasn't this union.

After coffee, Jergen and I are sent to the opposite side of the building where I hear a radio, but no—it's one of the carpenters, hanging by a safety belt, singing opera at the top of his lungs, hammering the whole time.

"That's Ben," Jergen says. "He's a little crazy."

I follow Jergen's steady plod to a concrete wall that separates us from a series of apartments being framed three stories below. I've never been so high.

Cautiously, holding tight to the metal scaffolding, I follow Jergen through an opening to inspect where we'll extend the wall. We're standing on two two-by-ten-inch planks, not at all like the sturdy platform on the street side. As we step back onto firm ground, Simon joins us with a tall, skinny kid he calls Eddy, who looks about fifteen and who, he says, will be our helper. While Jergen shows Eddy how he wants the plywood set up and drilled, Simon asks if I'm okay with heights.

"We don't make anyone work high if they're uncomfortable. There's lots of other work to do."

After he leaves, I try not to wonder if Simon brought Eddy over because he assumed I couldn't go out there. Later I'll find out that on this job, the First Aid guy has asthma, so they never ask him to climb scaffolds, and the rigger—the guy who directs the crane and ties and unties loads—is a pipsqueak who can't be half as strong as I am but whose small size and agility are perfect for rigging. For years, we women have been told we're not strong enough for jobs like this, and now I discover that a good foreman allows for physical differences. All this, plus double my wages. So far the union is looking just fine.

Each trade on this job has its own shack, so it isn't until my third day that I meet the ironworker foreman.

"I just wanted to shake the hand of the first woman carpenter I ever met."

"Apprentice," I reply, but he shakes my hand anyway.

"It's nice to know there's a woman carpenter on site," he says. "But there'll never be a woman ironworker." Suddenly the handshake feels like a slap.

A week later, I'm going for water at one of the big silver jugs on the deck, when someone yells loudly, "Fucking whore." It's Kim, who I worked with smoothly all day yesterday. Supposedly he's swearing at a board, but it's unnecessarily loud, and I know he means it for me. It's happened before. They work with you, then get mad at you for being competent.

And I am competent. I seem to know what I'm doing here. Jergen actually asks my opinion from time to time, and I notice Eddy, a first-year apprentice, making the mistakes I used to make. Perhaps this mini-burst of confidence is one reason I manage to come up with a good one-liner the next day. It's 7:30 a.m., pouring rain, and we're all standing in line, half-asleep, waiting our turn to go up the ladder, when suddenly the guy at the head of the line stops dead and we come to a soggy halt.

"Ladies first!" he says, gesturing at me, four or five people back.

It's ridiculous. An outsider couldn't even tell there was a woman under all this raingear. "No ladies here, just carpenters," I say.

One of the pleasant surprises is that—though I still don't like walking narrow top plates—working at heights doesn't bother me. We're always working off scaffolds at least ten feet high, often twenty or thirty feet, and though I'm acutely aware of where my feet are, I'm not afraid. In fact, as the walls grow under our hands, I have to hide my excitement. I'm plugged in, an essential part of this city, proud to be rising above the street storey by storey, breathing air that's never been breathed by other humans, only birds. As we rise past them I feel sorry for the office workers sealed up inside their air-conditioned offices.

"Are you married?" Jergen asks one afternoon.

He doesn't strike me as the kind of man for whom "living together" would be an acceptable answer, so I tell him yes. "And you?"

"To the finest woman in the world! You should know," he continues, "I'm a very traditional man. My wife has stayed home and raised our children and kept the house all our married life."

Suddenly it makes sense, why for the past few days he's been ordering me to "Stay there, you'll fall," whenever I start to walk out over heights, or saying "I'll carry that," when we have to move ply. I've been ignoring him, doing it myself anyway.

"I was raised always to protect a lady," he continues.

"But I'm a construction worker, Jergen. If you don't let me do my share of the work," I say, only half-joking, "they won't give me full pay."

I glance over at Eddy, but his focus on the job seems to say, I don't care if you're purple—I just want the paycheque.

"Don't you think ladies should be protected?" Jergen won't let it go. How to answer without insulting him or his wife?

"I think ladies who need it should be protected," I say weakly, "but I can look after myself." Then I think of something he might understand. "If I didn't work, my husband and I would never be able to afford our own house." And am absurdly grateful when, just then, the crane drops a pile of ply in front of us.

It comes up again when the new guy starts. The first thing he says when he sits down in the shack is, "We have girls now. I guess if women are carpenters, us men can stay home." Then, "The main thing I want to know is, how do you handle the weight?"

He's looking at me, but before I can answer, Idar, the new job steward, says, "Ask Eddy. Whenever Jergen wants something done he says, 'Ed, be a gentleman and carry this for me!'"

In this way, Idar sticks up for me. But the new guy proceeds to tell us all how, on his first day as a skinny runt in construction, they gave him a ninety-pound jackhammer and "I nearly died. Now," he says, "I can do it, no problem. I've learned how to handle the thing."

This is my cue. "Me too," I say, more cheerily than I feel. "You have to get good at balance and leverage," and am relieved to hear grunts of agreement around the table.

But something about me being here bothers them. One day I'm whistling to myself when a voice from the other side of the wall says, "It's not ladylike to whistle."

It's Herman, the carpenter who yesterday called me "baby."

"Herman," I say, poking my head over the wall to look down at him, "everything about me being here is unladylike!"

But the real shock is that, come payday, I actually feel that I'm worth the huge amount of money I've just been given—exactly like the men around me. Why not? I can see with my own eyes that I'm working at least as hard, producing at least as much as any other third-year apprentice on this job. Maybe it's this that makes me bolder, but that afternoon, when we ask our labourer for a ladder, the wooden one he brings is old and wobbly. One rung is cracked and tied with a bit of wire.

"That ladder's dangerous," a carpenter working nearby says.

"You wanted a ladder," the labourer mutters, "you got one."

When I stand uncertainly at the foot, Jergen pushes me aside. "Let's get some work done."

He goes up, but as he reaches to check a measurement, his weight shifts and the ladder begins a slide across the column's surface. I throw my weight against it long enough for him to rebalance, then brace the bottom until he's done. When he climbs down, I tell him we shouldn't be using it. He ignores me.

"If somebody accidentally cut one of those rungs," the first carpenter says quietly behind me, "that ladder would be toast."

Everyone carries on as if he hasn't spoken. As soon as Jergen's finished work on the column, I cut the third rung from the bottom and the ladder collapses.

"There's one ladder nobody's going to fall off," the first carpenter says to no one in particular.

A few mornings later, when I come into the lunch shack, I find a *Playboy* magazine on the first table. It makes my stomach ache. I sit down at the second table and try not to look, but until the bell goes, my ears strain for the crackle of glossy pages turning. By lunch, the magazine has moved to my table. It's like an elephant in the room. I tell myself I'm being too sensitive, but when I look up, Eddy is staring at me, the magazine open in front of him. He looks down at the centrefold, then back up at me.

Luckily, he's a first-year apprentice and I'm third-year. Holding tight to his eyes, I say, "Don't even think of it, Ed." I'm trembling, but the magazine is gone by the end of the day.

I can't quite relax, can't find my feet on this job. One afternoon, I'm working alone on the outside parapets when the site superintendent sets up a level near me. When I glance over, he's watching me. I keep working.

"Who makes the decisions in your house?"

My drill suspends itself in mid-air. "What do you mean?"

"When there's a decision to be made, is it your husband or you who makes it?"

"We both do," I say, and push harder to make the drill a bit noisier.

"I mean, an important decision. If you were buying a house, say, who would make the final decision?"

I stand up and face him. "We both make the money," I say. "We both make the decisions." He shrugs and moves away. Then Benny, the opera singer, asks if I'm married. Though I say yes, he begins to regularly bring me homemade wine and vegetables from his garden. In the lunchroom, when someone asks if I cook a hot supper every night, I answer absent-mindedly, "Only every second night. John does the rest." And I wonder if it's my imagination that the shack suddenly goes quiet. As if every man in the room is thinking, Thank God I don't live with her! As if I'm some kind of aberration. The contradictions exhaust me.

—

Is it around this time that I finally meet Emily Carr? No one who spends time in British Columbia can miss her paintings of heavy ever-green trees, Native totem poles and villages set in dense rainforest. I'd always found them dark and uninteresting until one day I see a particu-lar painting in a gallery and have to sit down, the wind taken right out of me. I've lived among trees, burned them for firewood, worked for several years in construction now, and I recognize here another woman who knows wood. On weekends I begin to hang out at the Vancouver Art Gallery, often writing as I sit in front of Carr's work. And I read. Emily was a splendid and prolific writer, and her journal, *Hundreds and Thou-sands*, becomes a kind of bible for me. Emily Carr, who lived from 1871 to 1945, had been told, like me, that women can't do certain kinds of work, can't be artists. But she did it anyway.

—

For my first union meeting, I pick a conservative shirt and blue jeans. Bill said I'd have to stand at the front for the members—all men—to see me, and I don't want to stick out. Though somehow I suspect that no matter what I wear, I stick out.

The meeting hall for Local 452 is smaller than I expected. A small stage with a table on it, a microphone and two chairs face a room full of about forty men talking and calling to each other, leaning over the backs of grey metal chairs. As I sign in and take a copy of the agenda, the room seems to grow quiet. I take a seat near the back, and talk swells again. Occasionally men glance at me, but always they look away before I can nod hello.

At 7:30, a man who introduces himself as John, president of the Lo-cal, calls the meeting to order.

"Brothers," he begins, then pauses, glancing toward the back, "... and sister..." He asks if there's anyone here, except the members to be initiated tonight, who isn't a member of this local. Some of the men look over their shoulders at me. A few of them smile.

When the president calls for correspondence, the secretary beside him reads out a series of letters, mostly requests for money from community groups, one from a union on strike. Most of it means little to me, so I sit back and enjoy the role reversals: a woman carpenter, a male secretary. Every report is preceded by an awkward "Brothers and, er... sister ...," which makes me feel both acknowledged and even more conspicuous.

I notice that the hands of the men nearest me are tanned, but white-skinned. When I double check, sure enough: there are one or two Asian men, another with darker colouring, but almost everyone here is Caucasian. Still, they have beautiful hands, swollen with muscle and maybe a touch of arthritis. These hands, I know, can handle the iron tools of persuasion—crowbar, hammer, screw jack—as easily as they can plane wood shavings fine enough to wrap your lunch in. There's a dynamic tension, a confidence in the hands around me, like springs coiled and waiting, eager, knowing what to do. My own hands are small and thin by comparison. There's dirt and form oil under my fingernails, and where I cut myself at B&D, the baby nail grows back crooked. But my hands too, are muscled. Working hands. They aren't what Elizabeth Arden would call beautiful, but I'm proud of them.

For the last item on the agenda, the president calls out the names of the proposed new union members, and I walk like a wooden thing to the front of the room to stand with six male apprentices.

"Is there any objection to any of these brothers, or sister, joining our union?" the president asks. There's a moment of silence, like that moment at weddings where the minister asks if anyone knows a reason these two cannot be joined. If one of these men says no, I'm out.

"Then I ask you all to welcome these new members to Local 452 of the United Brotherhood of Carpenters and Joiners of America," the president says, and grins. Everyone applauds. I'm in. I'm a sister in the Brotherhood.

—

August is a scorcher. Jergen prides himself on never sweating and never drinking water, but by the fourth day of temperatures over ninety, he's drinking water by nine in the morning and blaming me for the sunshine.

Accusations like this, spoken supposedly in fun, give me that creepy "witchy" feeling. I try to focus instead on how Jergen works always at the same slow and steady pace. I like his care and the way he takes nothing for granted. "Measure twice, cut once!" he says. There are no surprises with Jergen, no mistakes.

One habit of his gives me unexpected pleasure: Jergen calls everything "he." Usually in construction, everything except the tools is female. If someone wants you to shift something, the order is "Move 'er." If it's a very small distance, a guy might say "move 'er a cunt hair," though I try to forestall the "cunt hair" measurement when I see it coming. The studs aren't female, but since they're rigid and erect and do the heavy work of holding up the structure, nobody questions their gender. And tools, those big, noisy, steel-hard implements, are blatantly male. When a guy uses a tool with an extension, like a drill or a concrete vibrator, he carries himself with a conspicuous swagger, and we all know why. There are contests over, "How big is your hammer?" mostly meaning, "How heavy is the head? How long is your handle?"

The first time Jergen and I lift a sheet of ply into place and he says, "Nail him!" I think I must be hearing things. But no; he does it again. Perhaps it's because English is his second language, but I embrace Jergen's "mistake" with gusto.

"I'll nail him now!" I yell back, delightedly.

Until Simon, the foreman, overhears us.

"It's 'she,' not 'he.'"

"Actually," I say, "boards are 'it,' neutral."

But I know I'm onto something when, a few days later, there's a rush on and the superintendent assigns everyone to work on a single wall.

It's a gang form. Since snap ties aren't strong enough to resist the force of concrete in such a big wall, we have to prepare bolts that will hold the wall together. They're called she-bolts, three-foot-long ties like swords that will be threaded through to the other wall and secured with an oversized washer and nut. But she-bolts have to be greased so they can be removed after the concrete cures. I roll up my sleeves and begin slathering white jelly up and down the long shaft. I've heard of circle jerks but never, for obvious reasons, been able to participate. Now here I am with my union brothers in what feels like another initiation, having a good time, when the superintendent says to me roughly, "You better get back to what you were doing."

"Oh, no," I say sweetly. "I'm happy to help."

The superintendent paces, clearly in pain. Finally he grabs the bolt out of my hand.

"Go back to work!" he shouts. I tromp back—alone—to the wall I was working on before, smiling.

This is partly what leads to one of Women in Trades' first public awareness campaigns. At a meeting, Judy tells us her local lumber yard has given all its employees T-shirts with "It takes a stud to build a house" emblazoned across the front. After that, I start collecting terminology: the studs on a job are "erected," and there's a cement company whose logo is "Our product is always rigid." When you're fastening things to concrete, you worry about the "depth of penetration." Electricians "pull wire," plumbers have tools that include a "ballcock." A "nipple" on a worksite is any small thing that sticks out. We're always "screwing" things, and sheet metal workers, electricians and electronics technicians talk about "male" or "female" connections. Protrusions can even be called "male bosses." When I pull all this together in an article for a feminist magazine, even the kindest men I know tell me I've gone too far—but the tradeswomen love it.

One day, the ironworker foreman stops me and asks if I've seen the sign. "You better check it out," he says. His tone is flat, and instantly I feel fear. What if someone's scrawled something awful about me where the public can see? I head to floor level. They're pouring concrete on the outside walls and have posted the usual "No parking" signs, but today there's another sign, hand-painted. On a half-sheet of plywood, in letters one foot high, someone has written, "CAUTION: MEN AND WO -MAN WORKING ABOVE." I feel ridiculously proud. I exist!

Another day, Jergen and I head deep inside the belly of the building, where Simon has asked us to form up an afterthought of a wall. The job involves attaching eighty feet of two-by-four plates to the floor, using what they call a powder-activated gun. I can't wait to finish the job and get back outside; I know from school how dangerous this tool can be to your hearing. One of our instructors went deaf in one ear because of it. It doesn't sound that loud, he'd told us, especially in a closed space, because it works in a high-decibel range.

I tell Jergen I'll go for the hearing protection. A box of earplugs is always kept inside the First Aid shack where we can help ourselves.

"Don't bother."

"Do you already have a pair?"

"Get yourself some. I don't want any. I'm already deaf." Which I know isn't true, though many of the older tradesmen are hard of hearing. I bring back two sets of earplugs in case he's changed his mind.

He hasn't.

Later, one of the carpenters tells me that in the early days you could get fired for leaving a job to get something sissy like hearing protection. "So a lot of older guys still won't use it."

Regularly now, Jergen brings up his favourite line: "Ladies are angels and shouldn't do this work." At first I joke about it, but when he won't stop, I get serious.

"Jergen, I weigh one hundred forty-five pounds. If I ever fall off these beams you can be sure I'll fall as hard as you. I am not an angel."

"If you were my wife, you wouldn't be here."

"I'm not asking you to marry me, Jergen, just work with me!" I try not to sound as if I'm begging.

Lately, he's always angry. I try to foresee his every need, be the best apprentice he's ever had, but every night I come home increasingly bad-tempered. Nothing I do makes any difference; and it's getting worse. We're now three stories above street level and have fallen into a routine of setting up a forest of scaffolding and, on top of it, metal joists on top of which we'll lay plywood for the concrete. We're always working at heights, and Jergen's always talking about the danger.

"He makes me nervous," I tell John, but John has no patience.

"Don't take it so personally," he says over and over when I try to tell him what it's like at work.

"How can I not take it personally!" I say in a rising crescendo. "It's me—personally—he's talking to!"

But one night, after Kevin's in bed, we manage to slip between the spaces of taking it out on each other and start talking. I pour two glasses of brandy, two more, John lights the fire, and we talk. I go to bed at 2:00 a.m., and at 7:30 when I turn up for work, I'm still drunk. And I don't care what the fuck anybody thinks. I put my head down, ignore them all, feel nothing and build. All day, my cuts are fast and perfect, every time. Once, I catch Jergen staring at the end of the board I've just thrown into his arms, seconds after he asked for it. His mouth is open. I look down again at the wood—wood is my only friend—and keep going. It works better if I don't think.

You might suppose that after that, everything would be different; but no. I shrink back into caring what they think, and Jergen stays unhappy. Every night John says, "Don't take it personally. Use your anger," and I have no idea what he's talking about.

"Never think about falling," a carpenter once told me. "If you do, you'll fall for sure." But Jergen now yells, "Don't fall!" every time I step out onto a beam. It makes me nervous, and once I almost slip. I'm beginning to feel like I did on the B&D job, like a small animal, crouched, eyes wide, watching the larger animal's every move.

When we're sent to work on the walls surrounding the elevator shaft, Jergen walks over to Simon to get the details. Will it please or annoy him if I go stand beside him? Should I get things ready for when we start? I stay in the circle of sawhorses, arranging lumber, laying out cords. When Jergen comes back, he walks over to the two-by-fours I've laid out and begins to measure. Always before, he's immediately come to tell me what Simon wants—that's what partners do. My throat fills, but I push it back. He'll tell me in a minute. He's thinking, making some tricky calculation. But he continues to cut and lay two-by-four as if I'm not here.

"So what are we doing now?" I ask, forcing brightness. He says nothing. I move to his left to pick up a two-by-four, but he steps in front of me. Caught between him and the steel bars that mark the edge of the elevator shaft, I can't move. There's no reason for him to be here, blocking me.

"Jergen!"

He doesn't look at me, just stands frozen, ridiculous, letting me know beyond any shadow of a doubt he doesn't want me here.

After coffee, Simon sends everyone to get the next floor ready for the ironworkers, first thing tomorrow morning. I'm grateful there are so many of us; no one will notice that I work alone. We're twenty feet up, over what will one day be squash courts. Some of us are laying four-by-four wooden joists over the scaffolding Ed and I built last week, while others come behind, laying plywood.

"Watch your step here," Simon warns. "Some of these four-bys are short and not supported at both ends. Could be a quick ride if you step on one."

I replace the short beam I was laying with a longer one that's now supported on both ends—safe. When I look through the empty space, I can see the grey slab scattered with rubble two stories below. Rebar rods stand like swords, waiting to anchor another wall to the floor. Someone forgot to cover them. Those sharp ends would go right through you. But we're in a hurry, and sometimes small things like that are missed.

The dumb routine of laying beams is soothing. When we run out of long lengths, we start using shorter ones. It's dangerous, but it can't be helped; this is a rush. I remember seeing a few longer beams behind the sawyer's shack, and without waiting for a labourer, I walk over and pick up three. Usually I carry lumber on my shoulder, but this time, trying to hurry, I carry it in my arms, balanced across my lower body. As I walk back over the solid ply of the deck we've just finished, then across the exposed four-by-fours, I step on one of the ends that's not supported, and suddenly there's nothing under my feet.

I watch my right boot step into darkness and carry on straight down. There's a bit of metal showing through the leather toe; I always meant to get that fixed. My left foot follows into the hole. There's no sound. The air feels thick. The four-by-four I'd stepped on falls through darkness in slow motion below me. Less than three feet long—it should never have been used. As my body is sucked down, I have a snapshot of everyone carrying on around me, heedless. I'm invisible. I will now, finally, disappear.

Then I remember the beams in my arms. Through air thick as honey, I swing them at right angles. There's a loud *thunk* as wood catches on the beams on either side of me. I'm instantly suspended from the waist down in a narrow well of vile air, clinging to the beams splayed out in front of me like some fool's version of a life raft.

Now comes noise: shouts and the knock of wood and an echo of boots and the sharp smell of lumber. I'm shocked when I feel rough hands pull me up. When I look down, I catch a glimpse of steel bars, the ends sprayed red, for danger. The air is thinner again. I can breathe.

"Someone start laying plywood," Simon orders. "Those beams are dangerous."

That night, when I get home, I'm perfectly calm. Jergen has won. I've tried my best, but it hasn't worked. Nothing I can do will win me entry to this world. I won't be a carpenter after all. Then the phone rings, and it's Janet from Women in Trades. We hardly ever call each other between meetings.

"How are you doing?"

I cry as I tell her that I'm quitting, I can't do this.

A few minutes later, Heather, the boilermaker, calls. Then Judy, another carpenter. It's an extraordinary coincidence, and each woman gives me back another piece of my courage. Their sympathy is full of knowing. Over and over, they remind me that I can do this work. I am doing it.

"Talk to the union!" Heather says.

The next morning, when Jergen tells me yet again not to walk out on the beams and sends Eddy in my place, I know that if I don't settle this with him now, I might die here. I refuse to pick up the end of plywood he offers me, so he's forced to stop. When I tell him Ed is doing my job, he looks surprised, and I feel better already. If he doesn't let me do every part of this job, I continue, then I'll never learn how, and people will say women can't do this. It feels great to finally get this out in the open. I remind him that I'm a third-year apprentice, and if Eddy—a first-year—can do the job, then so can I. "And if I try, and can't do the work," I finish, "I should be fired."

"Don't be silly," Jergen says. He refuses to meet my eyes, though I stand directly in front of him, up close. "Eddy's a man."

He knows immediately he's made a boo-boo.

"I'm only worried about you," he says faintly. He bends down to inspect the drilled holes in the plywood.

"I want to do this, Jergen. I love this work. Let me do it."

"I don't want you to hurt yourself," he repeats stubbornly. "I'm only trying to protect you."

"You hurt me when you try to protect me!"

He comes out with a long speech that at first seems unrelated but is delivered with such passion that I listen carefully. He has three children, he says, "good children," who don't drink or smoke and lived at home into their twenties. His wife has never worked.

"I didn't stop her," he says. "She wanted to stay home. If my wife did this work I wouldn't know what to do."

And there we have it. Jergen is going through his own crisis. After a long pause he looks at me. "I guess we could see," he says. And on that vague note, we get back to work.

That night after work I go straight to the union hall to find Bill, the Apprentice Coordinator, who suggests changing partners. It doesn't happen often, he says, because nobody likes to rock the boat. Foremen usually notice what's going on and make changes before things go too far. Either Simon hasn't noticed, or he's waiting to see if Jergen and I can work it out. And I'd entirely forgotten about the job steward. Maybe he's waiting to see what happens, too.

The next morning I ask Jergen if he'd like to switch partners.

But Jergen's been doing some thinking of his own.

"If you were my wife...," he begins. When I open my mouth to object, he hushes me and carries on. "If you were my wife, I'd buy you some pretty panelling and some nice tools and let you fix up the rec room."

From Jergen, it's a huge concession. And things do get better. If he's still thinking I'll fall, he never says it out loud, and we work as partners until August, when I ask Simon to lay me off. There have been other layoffs, we're nearing the end of the job, and I want to go help my sister, who's just had a baby.

When I tell Jergen that this will be my last week, he's quiet for a minute.

"It must be hard, being the only woman."

We've come a long way from me being an angel.

At coffee break on Friday, Jergen shakes my hand and Ben, the opera singer, slaps me on the shoulder. No one else says a word, though they all know I'm leaving. When I go for my separation slip, the superintendent says, "Now that you're going, you don't really think this is women's work, do you?"

"Damned right it's women's work," I say as I take the slip and walk out the door. The empty space I'd crossed on my first day is now a patio with a set of stairs going down to the street. As I walk down them, the concrete balustrade feels cold and solid beneath my hand.

11

exotic dancing

In January, the union sends me to school for my third-year training. When some of us in the class go to the pub after school one Friday, one of the guys, Richard, sits beside me and yells over the din of the entertainment, "What side do you carry your hammer on?"

It's not the noise that makes me ask him to repeat it. He really is asking this fascinating question I'd never dare ask a man. Most right-handed carpenters carry their hammers on their left, which always seemed awkward to me. Turns out Richard and I are both right-handed and carry our hammers on the right.

Later, he leans over to yell in my ear, "What do you think of the Johnny-on-the-Spots?"

He's serious. I'm on my second beer.

"They all need a Tampax dispenser, and the seat's too high." We decide the latter is for capacity's sake. Then we talk about wages, bosses, the weight of hammers and future plans; we even talk about fear.

At first, he tells me, there were two or three times a day when he'd feel his balance go. "Now, even if I get that feeling, I know I'll catch myself. My reactions are quicker."

Richard is the fastest and most skilled student in the class. His love of the trade matches my own, and it seems to make not a whit of difference to him that I'm a woman. Suddenly I don't feel weird; I have a male carpenter friend.

On the day I finish my final third-year carpentry exam, I phone Dispatch at the union and ask for a finishing job. Steve tells me there are eight hundred people on the dispatch board, seven hundred of them want finishing, and he hasn't had a call for a finisher in two months.

"So what'll it be, formwork or framing?"

I take framing. I'm scared of walking the top plates, but Al stressed at school that every carpenter should know how to frame.

It's a three-storey walk-up, one floor of concrete and two of wood, in Vancouver's West End.

"You ever framed before?" the foreman asks in a voice bristling with doubt. When I say yes, he turns on his heel and stomps toward the office.

In the carpenter's shack, Marty, the job steward, glares at the rest of the crew and announces that it will be "fun" to have a woman around. No one says a word.

The foreman sends me to the second storey to lay joists with a man named Hugo. I'm awkward at first, but I soon remember the rhythms of the work and fall into packing plywood and two-by-fours, nailing and cutting—until I feel someone's eyes on me. A grey-haired man peers anxiously up from the first floor.

"What's your name? Are you a carpenter?"

"Apprentice. Third-year."

The man, who's the superintendent, soon appears on the deck where we're working. "Now, you look after her, Hugo!"

I'm relieved when Hugo says quietly, "She's doing fine."

Most of the guys here are younger than on the high-rise; one of the foremen looks about eighteen and someone says this is because "framing is a young man's job." Maybe it's because they're young that they don't seem as careful, especially about safety. Once, the foreman takes off fast when someone says the deck is sinking under a load of two-by-sixes, and another time, when the forklift doesn't push a load far enough onto the deck, the lumber falls off the edge of the building—fourteen-foot-long two-by-sixes falling two stories, to where a labourer had just been working below.

The foreman says, "We didn't kill anybody that time."

I have my own reminder of safety when, to speed things up, I tack a row of joists at one end, planning to go back later to finish nailing, but I forget one. When Gerry, one of the carpenters, steps on it and almost goes down with the joist he says quietly, "The life you save may be your own." After that, I get scrupulous about nailing.

In return for twenty dollars every payday, I get a ride to and from work with Gerry, who fills me in on job gossip. He tells me the owner of this company hasn't been able to come up with money for paycheques a few times, and one of the foremen does uppers all day and drinks all night—which maybe explains why safety's a problem.

Speaking of which, I'm starting to worry about the section where Hugo and I are building a rear stairway. The alley is lined with power cables, and high on the pole over our heads are two heavy-duty transformers. I only know they're called that because I ask Marty what two garbage cans are doing perched on a telephone pole.

The next morning Marty announces that Hydro has to move those power lines.

"It's illegal to put up walls within ten feet of that kind of line," he says, "let alone transformers, without written permission from the Workers' Compensation Board—the WCB—and even then, they should be protected."

Someone scoffs. "Right! Some bureaucrat signs something and then it's safe!"

"It means someone's made sure you're not going to get electrocuted, goof!"

"This is all bullshit!" someone says. "Just build the damn thing. Nothing's going to happen."

Hugo and I don't say anything, even though it's Hugo and I doing the work. When we go back on deck, I tell the foreman we'll keep building walls but we won't erect them until we get the okay from Marty. A few minutes later, a very large supervisor—I can tell by his white hard hat—stomps up with Marty in close pursuit. While they yell at each other, Hugo and I watch. I know nothing about electricity, and I'm hoping it's merely paranoid of me to notice that the two carpenters they're all— except Marty—so willing to throw up against heavy-duty power lines and transformers are the Indian and the girl.

I've joined the union's Apprenticeship Committee, where they've told us there's a section of the WCB Act—8:24—that gives workers the right to refuse dangerous work. I mention it to Hugo, who mutters that people can also make anonymous phone calls to the WCB. But we talk as if it's all theory, as if there aren't two garbage cans hanging above our heads that could kill us. Everything in construction is dangerous, and jobs are scarce. We keep an eye on the cans, and we keep working.

At the next union meeting I find out the union has put in an official complaint, and the business agent in charge of safety has told our foremen that under no circumstances should anyone be on the stairwell where Hugo and I are now raising walls, "not until the power lines are either covered or de-energized." So a few plywood sheets are tacked up around the canisters, and Hugo and I finish the job, close enough to the canisters that we could reach out and touch them.

Still, I like this crew. Fred, the foreman who had trouble accepting my dispatch slip on the first day, tells me that the day before I showed up, he'd had a run-in with the union and thought they were sending him a woman just to irritate him. As soon as I turned up, he'd phoned Steve.

"Your guy in Dispatch," he says, referring to Steve, "told me, 'You wanted a carpenter? You got one.' Said that if you didn't work out after a week, I could send you back. I wouldn't have believed it at first," he says, "but you're showing up some of the guys here." I'm not sure if that's good or not.

One day I straighten up when I hear several of the men hoot and whistle as a woman walks by. Tony, my partner, looks at me, does a little shuffle and goes back to work. The woman walks faster, staring straight ahead. I wait until she's just below me, then call, "Hi! How's it going?"

"Fine," she says, and looks up, smiling. "How about you?" The guys' mouths are hanging open. None of them has ever worked with a woman before.

"See?" I say to them. "That's how it's done."

I know they're struggling to find a category for me. Soon after I started here, one guy made a pass at me and I let him know I wasn't interested. A few days later he came toward me, waving his little finger.

"I cut it." It was a minor scrape.

"You want me to kiss it better? Go to First Aid!"

But a few days later he approached again. "I don't get it," he started. "You're not my girlfriend and you're not my mother. You're my... my..." And as he fumbled for a word, I remembered that this is a union job.

"Your sister?"

He beamed. "Yes! You're my sister!"

I feel generally accepted by this crew, so it's odd when someone tells me in the lunch shack one day that Morris, the owner of the company, wants me to get a haircut. Morris is a shadowy figure I've seen only from a distance as he paces the sidewalk. There are ongoing rumours he's running out of money and wants us to work faster, but bosses always want you to work faster, so we ignore the talk.

"Morris thinks you should cut your hair," the guy repeats and someone snickers. A small warning bell goes off.

"Cut my hair?"

"Yeah, he thinks you're a guy." And everybody guffaws.

This is the first construction job I've been on, ever, where we have real conversations. Mostly they happen at lunch, about food, and mostly they're initiated by Gerry and me. We compare bean sprouts with alfalfa and discuss the best way to tell a perfect avocado. We talk vitamin supplements. We even dare mention the body. Gerry is a bodybuilder, the guy we all call when an especially heavy wall needs lifting. One day, I let it slip that I've been to a massage therapist. Instead of the shocked silence I expected, Gerry says, "Good idea. I should go to mine," and we're off on a discussion of massage and how to keep your skin from cracking in wet weather.

"Vaseline," says one guy.

"Nivea," another.

"I stretch every morning," Gerry says, "and after work, do a few chin-ups. You?"

I've never done a chin-up in my life, but after that, every time I spot a two-by-four at the right height, I try a chin-up. I never get higher than my forehead. It's those pectorals again.

I love these guys, and it seems mutual. In March, when I tell Gerry on the way to work that it's my birthday today, he tells everyone in the shack, and the crew sing "Happy Birthday" and shake my hand. I don't think it's just because I'm now thirty-five and one of the oldest ones on the job that they're happy to help me celebrate. Still, I'm careful. On the day I wear a T-shirt a little tighter than usual, they're oddly silent, and polite—too polite—handing me tools, running to fetch me nails when I mutter I need some. I go back to loose shirts.

I love the exuberance of this crew, their energy. And I'm getting better at "one-upping" along with them. When someone reports that the crew next door asked how I was doing, I say, "Funny, they never ask me how you guys are doing," and the crew laugh. This, I know, makes me seem "okay." It's how you're expected to talk—in one-liners, joking. I'm learning their language. When one of the foremen says, "son-of-a-bitch," every second word, as in, "You can lay out that son-of-a-bitch wall next," as a simple instruction, I actually find it charming.

Then again, maybe it isn't exactly the language I love; it's what it's a symptom of. I love talking with women, but our talk is at length, looping, circling, with lots of detail, context, background. These men are condensed. Everything—words, feelings, action—steeps inside them so it comes out espresso. Their physical need may be for women—I serious-ly doubt there are many gay guys in construction—but their romance is fishing trips and scoring the winning goal. Even if they don't actually do these things, they get satisfaction from imagining them—together. Men have the power of knowing who they are together. I begin to think their mystery doesn't lie in their physical strength or confidence, so much as in their brotherhood. Though we feminists talk about "sisterhood," it's a weak tea compared to the men's potent, practised brew.

These are the best days I've had in construction, and yet on some days I feel desperately foreign. When I ask one man outright why he's giving me a hard time, he says, "It's nothing personal; just that you're a woman."

And yet I like the fact that I'm different. I like the spark of the physical between the men and me. On the good crews, like this one, I allow myself to relax, sink into my body and just enjoy. Like my tool belt. Even after I've sewn it to its smallest size, it's too big and sits low on my hips. When I walk, if I swing my hips just a little—hammer on the right, combination square and nail puller on the left—my tools make a small music.

Six months into the job, when I'm one of the most experienced here, one of the carpenters says at lunch, "Kate, why don't you do an ex-otic dance for us?"

I snap, "Do your own exotic dance," but Gerry throws back his head and laughs.

"Kate does an exotic dance every time she walks down the hall."

He'd noticed. It gives me an idea—or perhaps it gives me permis-sion. That night I sew different colours of embroidery thread around the front pocket of my coveralls. No one at work says a word. Then I find a treasure at the Sally Ann: a plaid jacket like the men's blue and red ones, but pink. The pink jacket and the embroidery cheer me up. I've declared my gender, and it surprises me that the men treat me better than ever. Perhaps they'd been confused by my trying to hide. Now that it's out in the open, they can relax a little.

And yet... I'm like a kid looking in the window at all the men gath-ered inside around the stove, sharing a drink and a laugh. I do as they do. I make myself inconspicuous (or I try the opposite tactic and wear pink). But the closer I get to "my" crew, the more impossible it feels to cross the last tiny gap. I press my nose to the glass, "Let me in!" and on some days they hear me, wave, "Sure, come on in!" but they don't know how to open the door, don't have the key. Or is it that I don't know how to turn the handle?

One night as I'm falling asleep I have a moment of feeling extraordinary power, utterly right in my bones exactly as I am. It passes in an instant, but I know I've just had a glimpse of the sense of confidence that most men—at least, most construction workers—take for granted. As a woman, I've never felt it before now.

—

One morning a new foreman tells me he needs me upstairs to lay decking with Cal. Maybe it's because this guy's new that I have the courage to say no. I've never said no to a foreman, but Cal is different. I can't figure him out. He's a loner, for one thing. Unlike everyone else, he works by himself and rarely eats in the shack with the rest of us. On the odd time he does, he only joins the conversation to make cynical comments. I don't like him. The others call him a pig and a wood butcher but put up with him because, they say, he's the best framer here. But they don't like him either. And he makes no secret of the fact he doesn't like women. Once, when I was working near him, he moved a large bundle wrapped in green garbage bags out of my way.

"My chainsaw," he said. "She's like a woman. I sleep with her, only I have to leave the bag on because she smells so bad."

No, I don't want to work with this guy. I don't like him and he doesn't like me.

"I'll make you a deal."

I've never seen a foreman so polite. He's short of carpenters and we can't start anything new until the roof is on, so he asks me to work with Cal just for today, and if it's too bad, he promises to bring me back downstairs tomorrow.

When I step out on the roof, Cal is starting the bridging, nailing two thin pieces of wood in an X between joists to convert the floor into a locked grid. Given that this job takes up a city block, I can do my share and still stay out of Cal's way, which suits me fine. I set up one two-by-ten for my materials, and another to sit on, then start to nail. It's easy work and soon I'm enjoying the sunshine and the peacefulness of being left alone, happily setting two two-inch nails into the top and bottom of each piece of bridging before I nail them along the chalk line Cal marked.

"Don't bother with two nails," a voice says.

He's standing over me.

"The Building Code says two nails each end."

"Waste of time," he growls, and goes back to his side of the deck. He's already halfway along one row and I've barely begun.

"You're going to have to redo yours!" I call, knowing the only reason I can be so cheeky with a journeyman is because I know I'm right. School drilled us on the Code.

"Good enough for the girls I go with," he replies and I shut up. This isn't my problem.

I still have most of the final row to nail when Cal begins to lay plywood. This isn't a sloped roof, where you need a gap between sheets, like I've done before. This roof will also be a deck, so it's tongue-and-groove sheathing, which means a thin "tongue" at one side of the ply must be fitted tightly into the groove of the other. Cal lays the first row, moving quickly. And talking. I try not to listen but he's complaining about how dull construction is. I can't resist.

"I don't find it dull at all," I tell him. "I like it. And when I'm not doing construction there are lots of other interesting things to do." I tell him I write articles for women about being in trades, then laugh, partly because it doesn't sound like it would be the least bit interesting to Cal, and partly because I've never revealed myself this much before on a job. There's something about this guy that I take as a challenge. He's like my dad in a blue collar.

After a few minutes he says, "Are you a feminist?"

I look up, measuring him. "I don't usually tell people this, but yes."

"Do you know what I am? I'm a male chauvinist pig!"

"Oh yeah? I didn't think you were." I'm being facetious, but he takes me seriously and—I'm surprised—it pleases him. A few minutes later, when we're working close to each other again, he says, "You know why women don't become tradespeople?"

He said "women" and "tradespeople," not "girls" and "tradesmen."

When I don't answer, he says, "Women can't be tradespeople because of their own minds. They don't believe enough in themselves."

Now how the hell has he figured that out?

"Come here," he says. "You want to see how it's done?"

I watch as he holds the sheet of plywood vertically, groove side down, against the tongue of the first piece, then lets it drop away from him. As it falls, he places his boot on it and guides it to come to rest where he wants it, snug against the tongue of the first piece. I can't help myself: I grin. It's a neat trick.

"You try it."

The trick is in your foot. When I botch the first sheet, he shows me again, using his boot like a hand to guide the falling sheet to exactly where he wants it. When I drop my third one perfectly into place, Cal grunts approval.

Cautiously, I begin to work with him. As each sheet is laid we take turns standing on it, holding it down while the other slides the tongue into the groove with a few blows from the sledgehammer, protecting the delicate lip with a piece of two-by-four. I feel vaguely uncomfortable at the tongue slipping into the groove, but Cal says nothing, and soon we have a whole row done and tacked. Another.

Several times Morris comes by, turning his back on me and having terse conversations with Cal that end with Cal saying he has to get back to work.

"What's he want?" I finally ask.

"Wants to know how you're doing."

I must be feeling brave. "So?"

"So, I told him I was making a framer out of you." He stops and looks at me. "You're doing fine," he says.

At the end of the week, when the foreman asks if I want to switch partners, I say no.

One day, Cal and I are framing around a cement block firewall where Code calls for a two-inch open space all around. My job is to nail a joist, on edge, onto the plate against the firewall, and since two inches isn't enough to get my hammer into the gap, it has to be nailed from this side. My first impulse is to toenail into the joist, but it's obvious even before I start that the nail will go into a thin sliver of plate and then air— not enough to hold.

Cal notices. He kneels beside me and gently bends a spike over the edge of a scrap of wood until it forms an arc, then sets the bent nail, curve toward us, into the edge of the joist—as I had, but mine was a straight nail. His nail bites into the joist, then follows its own curve to return to the top plate. He drives it home, then tests the joist—solid. I look at him in wonder. So simple.

"That's beautiful."

He only grunts, but I can tell he's pleased. "Try it."

Still, working with Cal is a roller coaster: one moment of beauty, then a stunner out of left field.

"You want to know why employers don't hire more women?" he asks one day.

I'm cautious, but curious. Cal isn't mean, just heart-thumpingly di-rect. He's like a window for me into what the men are thinking.

"Why don't employers hire more women, Cal?"

"Because they distract the workers. I just made two mistakes."

"That's my fault?"

"Yeah. I'm aware of you."

But he also teaches me. One day he challenges me to cut a four-foot sheet of ply, by eye, without a chalk line to mark the cut. He laughs at my wobbly cut, but after a few days, my eye is true. I now cut two-by-sixes and two-by-tens without needing a straight edge, and two-by-fours two at a time to save having to measure a second time. He sets me exercises like ripping the length of a board with just one mark, using my thumb as guide, and walking with the nail gun bouncing in my hand so I can nail a floor as fast as I can walk across it. Most carpenters look down on framing as unimportant because it doesn't show, but "if the framer's work isn't done well," Cal says, "no one else's work goes easily. Ask the guy who has to hang the doors."

These days, I love the rhythms of my work, the outdoors zest of it, the moment of walking over to a pile of studs damp from the mill, the small bite of pitch to my nose as I make a smooth lift of six raw studs, their reassuring heft as I walk, anchored securely by the load on my shoulder. The word "raw" is made for new-cut lumber. The smell and feel of it give me the sense of a bond, as if I once knew this wood, as if the currents of blood and sap are the same. This is also the job where my tools finally become just tools—a means to do something—rather than weights I have to lug around. I use them without thought, as extensions of my arm. So simple—what is a hammer but a rock with a handle on it? Companion.

One day, I'm vaguely aware of Cal watching as I bend over, nailing. My hammer never stops its steady swing. It's a mark of the trust I have in him by now that it never crosses my mind he might have been just watching my ass in the air.

"You're starting to look like a framer," he says. It's the nicest thing anyone ever said to me.

—

When the roof is finished, Morris buys beer and snacks for a roof-ing party. Gerry explains that carpenters always celebrate getting the roof on because it means the basic structure is done. In the old days, they used to tack a small tree to the peak, something about appeasing the gods of the tree. But when the men begin peeling beer caps off with their hammers, I suddenly feel very feminine, as if I've accidentally wandered into a locker room after the winning game. I take the beer that's handed me and hug the shack walls when someone suggests a nailing contest. I like the competition of getting a job done, not the blatant testosterone of competition for its own sake. I squeeze closer to the corner, where Gerry and I get into conversation. After I've been sweating all day, the cold beer goes down fast. Someone hands me another that goes down almost as fast again, and suddenly I'm drunk.

"Going so soon?" Marty smiles as I push past him. "You think we'll see too much after you've had a few beer." He says it kindly. I'm already halfway to the ladder.

Shortly after the roofing party, layoff rumours begin. Sure enough, just before coffee break the following Friday, one of the foremen comes up to me, saying, "I have bad news." One hour's pay in lieu of notice— and I head to the shack to pick up my tools. Gerry offers me his hand to shake goodbye, then leans over and kisses my cheek.

"What the heck, eh?" he says, and grins.

Two of us are laid off, an older carpenter and me. As we trudge off the job, I ask him, "Do you ever get used to this?"

"Never."

The man drives me and my tools home, where I cry for a while before going to the hall to sign up again for dispatch.

"What are you doing signing that book?"

It turns out there has to be a certain ratio of apprentices to journeymen on every job, and Morris doesn't have enough apprentices. Steve disappears into the back where I hear him talking to a business agent, and on Monday morning, I'm back at work as if nothing's happened.

But that weekend, at a party, I run into a man I briefly worked with on my first union job, and we immediately get to talking construction. He tells me how being a construction worker gives men pride.

"We feel strong, proud that we can do this dangerous, important work, and at night we go home to wives and girlfriends who show us the tender, feeling side of being alive."

"So when a woman walks onto the job...?" I ask, knowing the answer.

"It's confusing." What he seems to be struggling with is that, "If a woman can do the emotional work at home, and have babies, and now do men's work as well, then what good...?"

He looks at me with a question mark, as if he can't believe what he's about to say. I can't believe it either. I finish it for him, to make sure.

"...then what do we need men for?"

"Yeah."

I go home thoroughly depressed.

—

Cal is now a foreman, and at lunch he tells us that the WCB and BC Hydro both turned up yesterday, at last. Effective immediately, they've closed the entire section where Hugo and I were working, and forbidden anyone to go within ten feet. One of the wires, they said, was an exposed twenty-thousand-volt line. It could have arced and done us in.

I get tears in my eyes. Someone in the shack says, "In the end, you didn't get hurt, did you?"

But we could have died.

—

On every construction job I've been on before, the foreman tells you what to do, leaves you alone to do it, and when you're finished, gives you the next thing to do. But supervision on this job is getting tighter all the time. Morris and Gordie, the new foreman, seem to choose one carpenter per day to hover over, questioning everything. The guys are starting to complain and soon it's all we talk about in the shack.

"Morris's hanging over me worse than my mother," someone says at lunch. "Today he was on to me for dropping nails, for Christ's sake!"

The next morning, a few of us are figuring out which decks get sheathed and which don't when Morris starts in again from below, "Five of you standing there for ten minutes, wasting time! You've done dozens of these decks, you should know by now!"

I happen to be feeling calm at that moment and say, "Morris, it's been less than two minutes, there are three of us and none of us has sheeted this kind of deck before."

Strange things happen to Morris's face and I worry he's having a heart attack. While he turns red, screaming, I start laying down plates. Where did everybody go? Next thing I know, Morris is standing over me, then Gordie, and now, of course, I start making stupid mistakes as Morris starts in again.

"Gordie, look at this. It's a mess. Get someone else over here..."

"Shut the fuck up!"

That voice? It's mine, screaming at the owner of the company I work for as I fumble furiously to unbuckle my tool belt. "Get off our Jesus backs!" I throw my hammer as hard as I can, with just enough restraint to heave it close but not directly at him.

Morris goes into heart attack mode again but clings enough to being a gentleman to pick up my hammer and hand it back, saying very formally, "If you can't take orders around here, then you can leave."

"I can hardly wait!" And I stalk off the platform and down to the first floor.

My partner's eyebrows are about to fly off his forehead and Gordie's mouth is going like a drowning thing, but not a sound comes out.

I won't let them see me cry. I'll collect my gear, walk off this job and go home. No, that's silly; I need the work. I'll go cry in the Johnny-on-the-Spot. But the Johnny-on-the-Spot is busy, and besides, I remember how it stinks and the graffiti is awful; I can't go there, either. Still fuelled by fury, I spot a two-by-four nailed across a door frame, six feet up. I grab it and lift myself. All the way. A full chin-up! I'm so startled that I forget Morris, forget about walking off the job. I did it! Ten minutes later I go back upstairs, pick up my tool belt and go back to work.

Morris has disappeared. No one speaks. The eerie silence is broken only by the sound of hammers and a load of studs being dropped on the deck. One guy whispers, "What happened?" but no one answers. Even in the shack later, no one speaks. From that day on, Morris is rarely seen except on paydays. He ignores the fact he's sort of fired me and I ignore the fact that I sort of quit. But for two days, the crew won't talk to me. I'm confused; aren't we all glad I got them off our backs? I tell myself I don't care. Why didn't they speak up? Where was the job steward? They're a bunch of wimps. And I did my first chin-up!

When I go to my books for comfort, the American feminist Adrienne Rich says, "Unanalyzed pain leads us to numbness, subservience, or to random and ineffectual bursts of violence.... Most women have not even been able to touch this anger, except to drive it inward like a rusted nail."

At Women in Trades, Janet thinks their silence is because they were shocked at a woman asserting herself, though I'm pretty sure throwing your hammer at the boss is not a textbook case of "assertiveness."

It's Gerry who explains it a few days later as he drives me home.

"You were mad," he says carefully.

"Of course I was mad, we all were! But I was mad at Morris, not at you guys."

"Yes, but you were really mad."

So that's it; I scared them. John always says, "Use your anger," so didn't I do just that? I was mad so I got mad. Isn't that the clean, honest way to act? I decide to hell with all these men. It's just too bad if they can't handle a woman's anger.

One week later we're all tying on our tool belts when Morris calls Gordie into the office, and a few minutes later Gord sticks his head out the window with a strange look on his face.

"Wrap it up, guys. The job is closing down." We're all laid off. Morris has run out of money.

⚊

I get a few more short jobs, but construction work is scarce. I'm eligible for unemployment insurance and I keep busy with a rush of invitations, through Women in Trades, to sit on committees, give speeches and make presentations about what it's like to be a woman in construction.

One of them is an invitation to speak to a conference on apprenticeship organized by the provincial Ministry of Labour. Tradesmen are getting older—their average age is now forty-eight—and there aren't enough apprentices being trained to replace them. It's struck someone that they could broaden the pool of apprentices by including the other 50 percent of the population—women.

I agree to give the talk, but I'm scared. These are people with the power to make important changes in apprenticeship. My dad, who by now is a Very Important Person in his own company, advises me to "think of them as a bunch of guys like me. They're just somebody's dad."

The older I get, the more my father surprises me.

But there's another problem. I'm damned if I'm going to turn up at the conference dressed in blue jeans and perpetuate their image of tradeswomen as hammer-totin', muscle-bound mammas. But the only shoes I own are boots, one pair of steel-toes for work, one pair of brown leather for everything else, and my wardrobe isn't much better.

John comes to my rescue. He phones his mother, and after an evening in her closet, I come out with a spiffy two-piece suit that's only a little snug. I buy the first pair of high heels I've owned in thirteen years and off I teeter to the opening dinner of the conference, where I stay seated as much as possible. They put me beside a high-ranking official who—knowing I'm a third-year carpenter apprentice—tells me flatly, "Women can't do this work." My speech the next day feels like a whisper in a hurricane, but I give it anyway, clutching the podium for balance.

But when Bill Zander, president of the Carpenters' Union Provincial Council and a member of the New Westminster Local 1251, asks me to speak to the carpenters' convention, I feel suddenly way out of my depth. It's one thing to offend government bureaucrats, but it could be dangerous to offend the men I work with. Still, Bill persists: if women like me don't speak up, who will? I make the speech.

—

One good thing about being unemployed is that I have time to read—right now, *Woman Hating* by Andrea Dworkin and *The Cinderella Complex: Women's Hidden Fear of Independence* by Colette Dowling. I also immerse myself in Aritha van Herk's novel *The Tent Peg*, about a woman who disguises herself as a man to get a cook's job in a Yukon mining camp. Partway through the book, the cook—who has by now been outed as a woman—has an encounter with a grizzly bear that ends with her sweeping off her hat in a deep bow and the bear leaving without doing her any harm. Two men happen to see it, and one gasps to the other, "She's a witch!" When I read that, I burst out laughing. These days I yearn for poetry, too. Maybe it's the ambiguity I love, or the beauty. Poetry doesn't fool around. It tells the truth without regard to gender or colour or power dynamic. My favourite poet right now is Adrienne Rich, who writes, "This is the oppressor's language, yet I need it to talk to you."

—

When the head of the women's committee of the BC Federation of Labour arranges for first a discussion paper, then a whole conference on sexual harassment in the workplace, there's much skepticism and grumbling in the labour movement, from women as well as men. "Much ado about nothing," people say, and "Women are losing their sense of humour."

I speak on behalf of Women in Trades, and we're astonished at how easily the women who work in offices understand what we're talking about. We all laugh a lot—with relief, because now we know what we're saying no to. Finally.

Women in Trades follows up with our own workshop on sexual harassment, where we talk not just about harassment but this other new term, "assertiveness." That's when you don't get mad, you just say clearly what you want. In the trades, I'm slowly coming to understand, it's not uncommon for a man to push you in a way that feels like harassment when he's just testing to see where your boundaries are. So you tell him—sweetly—"Fuck off." With a smile. Marcia, another carpenter, says you can get away with anything if you smile.

There's one part of me that sees all this, intellectually. It's what John's been saying for years. The other part of me—that part that comes down to actually being assertive on the job—seems blind.

In November, still with no union work, I get sent to school for my fourth and final year of training. Richard's still working and I know no one else in the class. When the teacher walks in, saying, "Good morning, men," I follow it up weakly with, "And women." But that afternoon I can't do anything right in shop, and on my way home, I start to cry. I'm nursing his comment like another rusty nail.

"Don't let them get to you!" Marcia orders me on the phone, and tells me about affirmations. By Monday, John has to lift several sticky notes ("I have a right to do this job!" "I am a skilled and confident apprentice!") so he can get to the grocery list on the fridge, but the affirmations help. So does the joke Marcia gives me to tell.

On Monday, when the teacher rolls in again with "Good morning, men!" I say loudly, "Want to know why women don't make good carpenters?"

He freezes.

Feeling outrageous, I hold my fingers three inches apart. "Because we were always taught this is six inches."

There are a few nervous giggles from the class, and then we get to work. The teacher never calls us "men" again. But a few days later, when the men start bragging about all the highball crews they've been on, one crew sounds familiar. It's B&D Framing.

"I worked for them!" I tell the guy who's bragging. "The year after you."

"They got a lot slacker after I left."

Of course. How could a girl possibly keep up?

One afternoon, our class joins another to watch a video on worksite safety. It features a woman dressed at first in a pink jumpsuit and pink hardhat spackled with flowers, then naked from the thighs down, picking her way in bare feet over glass and rusty nails. I can hardly believe my eyes. In the final scene, she holds out a bandaged hand, hurt when she stupidly put it between a wall and a hard object. A man slips a wedding ring on her finger as the final title scrolls by: "Avoid traps."

I talk to the two women in the other class and we agree we'll write a letter of protest, but first we need the name of the video. When I go to the teacher's office to get it, he's soon shouting at me that he went to a lot of trouble to get that film, that sex makes guys pay attention.

"There must be a way to make us pay attention without using stupid women."

"We could use queers!"

When I join the rest of my class in the shop, I'm still furious. I desperately want Rob, the carpenter I'm paired with, to ask me what's the matter, but he stays busy building a complex set of rafters. Finally, I ask him outright what he thought of the video.

"Pretty bad." He keeps nailing.

That takes some of the wind out of my sails. I'd assumed the guys would agree with the teacher. But I'm too angry to stop, and off I go on a rant about how men and women will never be able to work together with that kind of stupid stereotype thrown in our faces, and if that teacher thinks he can drive me out...

When I'm finished, Rob puts down his hammer.

"Kate," he says patiently, "a guy who likes that film is a guy with old ideas. To you, he's sexist, and to me he's macho. I don't like it either, but the problem isn't you, it's him." Then he picks up his hammer and goes back to work.

I'm speechless. It sounds so simple. Is this what John means by "Don't take it personally?"

After that, things smooth out and I become good friends with a couple of the guys in my class. In December, I pass both my fourth-year and my Red Seal exams, which makes me a Red Seal carpenter, qualified to work anywhere in Canada except Quebec. Now all I need to actually get my ticket as a journeywoman carpenter are two thousand more hours of work.

I drop into the union hall to see if anything's happening. Up to now, heavy construction in the province has been done by union companies, but there are rumours that British Columbia's new Social Credit government is out to bust the construction unions—and it seems to be working. I head to the Dispatch Board, a wooden frame that takes up half the wall under a sign that says "Journeymen." This is the heart of the union. Members' names are written on what look like library cards, and the cards move slowly up the board as the people at the top are dispatched for work. I hang around for a while as men trickle in. An older carpenter, drinking coffee at the table beside the dispatch wicket, watches me search for my name, finally finding it at the bottom of the board.

"At least you're not in the shoebox."

There are so many unemployed journeymen, he says, that we can't all fit on the board, so Steve in Dispatch sticks the extras in a shoebox.

The carpenter's name is Ivan and he's been out of work for two years, picking up odd jobs. Even at that, he figures he's luckier than most. He has a savings account, his house is paid for, and his wife has gone back to working full-time as a practical nurse.

"I've been in this trade forty years, and I've never seen it this bad."

When Bill wanders in to say the Unemployment Committee is about to meet, Ivan and I decide to sit in, and soon we're both regulars at the committee meetings. One morning, some of us stand on a curb at rush hour to hold up signs demanding government action on unemployment. The adrenalin of it makes me feel more alive than I've felt in weeks. Another day we stand in front of a liquor store and ask people to sign a petition. But most of the time, the members of the Unemployment Committee try to ignore the feeling that there's not a lot we can do. It's bad out there. Carpenters are starting to turn in their union cards and go to work for non-union companies, taking their skills with them. What else can they do?

The worst part of unemployment is the feeling of powerlessness, so when someone phones to say the union is holding a demonstration, I go. It's at a job site where, for the first time in the province, a non-union contractor has been bonded so he can take on—in this case, take over— a high-rise project. The contractor's name is Kerkhoff, and the job is an apartment complex called Pennyfarthing, near False Creek in Vancouver.

It's a raw site—a huge hole in the ground surrounded by a chain-link fence. Someone hands me a hand-lettered sign that says "This is Union Work," and I walk in a tight circle with six or seven others. I come back the next day, and the day after that. And an amazing thing happens. Every day there are more people, more signs, and then it mushrooms, and within a week there are dozens of us. It's cold and wet as we walk, and someone brings two fire barrels and firewood to keep us warm. A coffee truck begins to drop by; someone says the driver is a Teamster who supports our picket line. That's how I find out I'm on a picket line.

There are carpenters on the line, of course, and a few men from other building trades, along with a growing number of signs that say "CUPE" and "HEU" and other initials I've never seen. One day, Heather from Women in Trades arrives. She's a boilermaker and I worry she's been laid off. But no.

"They told us at the hall you needed reinforcements down here, so we all took the day off." I get tears in my eyes.

As activity increases on the building site below us, the picket line gets more energetic.

"Scabs!" people yell at the men who drive through our lines. "Give us back our jobs!"

"We want work!" I call weakly to the men moving around on the footings below, then look around to be sure no one's watching. It's not that I don't want the guys in the hole to have work; I just want us to have some, too. Ivan points out video cameras in several apartment windows nearby.

"Smile. You're in the files now."

Which is how I know they aren't TV cameras, though the news cameras show up soon after, mostly at start-up and quitting times, when there's the most pushing and shoving and hammering of fists on the trucks crossing the line. I don't like the violence, but I keep going, keep wearing my sign, and as I watch through the fence, day after day, men doing the work I might have done, I feel a small knot of anger ball up tight in my chest. I shout louder. I forget about the cameras.

One afternoon, the non-union foreman below comes close to the fence and announces he'll hire anyone who wants a job. Someone on our side asks how much he's paying.

"I can't feed my kids on five bucks an hour," a guy on the line yells, and spits.

TV cameras record the confrontation and I watch it on the six o'clock news. On the picket line I felt fellowship, pride and a growing sense of purpose. On TV it looks meaningless and violent, not at all like the place I'd been that morning.

In the third week, one of the non-union carpenters comes up to the fence from his side and starts yelling at us, "Get a job!" I hear myself growling at him, furious, pressed against the wire fence. For an instant I lose myself in fury, and then the man falls, his head bleeding.

I've killed him.

But no, someone else has thrown a stone. The man stands up and, when he sees the blood, shakes his fist, turns and stumbles back to the shelter of the trailers on site.

"Serves you right!" someone shouts from our side. But as soon as the man fell, I'd felt the hush.

I've begun to make friends on the line, including with a woman in a blue jacket that reads "Hospital Employees' Union," who one day buys me a coffee. Bill Zander, the president of the Carpenters' Council, is always talking about "union solidarity," and this must be it, like a firm hand when you're drowning. But the next morning, as I round the corner to the site, something's wrong. Where yesterday there'd been more than fifty people, today there are only a few—and two men who talk to everyone who approaches. One of them comes toward me.

"Go home."

I look over his shoulder, assuming he's from the non-union company, but when I try to walk around him I see he's wearing a union jacket with a building trades logo. I stop.

"Kerkhoff has taken us to court," he says brusquely. People might go to jail, he tells me, get big fines if we stay here. The executive has ordered us all to go home "while the lawyers fight it out."

This can't be right. The executive has said at union meetings and on TV, over and over, that we won't stop even if it means jail for them. This issue is too important. And what do lawyers know about what's going on here, what it means to us?

"Go home," the man repeats.

Eventually we all do.

—

"Bitter" doesn't describe the mood at my next Local 452 union meeting. Guys line up ten-deep at the mike to speak. A lot of them don't wait.

"You said a fight to the finish!" a man behind me yells from his seat. "You said jail if you had to, we'd pay any fine!"

"You'd think it was your own personal money, not our dues," guys shout, and "We're letting the lawyers run this union!"

"Whose union is this, anyway?" one man roars. Someone else calls back, "Ours! It's ours!"

Finally the president bangs his gavel over and over, says if we can't be civil and talk one at a time then people are going to be ejected. The talk gets a bit quieter, and guys go back to taking turns at the mike. If that's all it takes to scare the union off, the consensus seems to be, then this union isn't the place they belong. The question hanging over us all is this: if this isn't the place we can have pride in skilled, well-paid work, then where is?

Over the next several months, more and more men start doing their own small jobs. Some drift off to the rapidly growing non-union sites. It's partly out of loyalty to the union that I don't look for other work. I try not to think about what I'll do when the UI runs out.

—

The only good news these days is that John has been accepted to law school, something he's wanted for a long time. And sure enough, after he starts school in the fall, he's in a better mood—when he's home. Mostly he's at school, studying. When he's not doing that, he's in the garage rebuilding a battered wooden sailboat.

"An Enterprise!" he announces proudly, as I stare at the battered hull.

"Do you know how to sail it?" I ask. A practical question, I'd thought, but it pisses him off. Everything I say these days pisses him off.

"I'll learn," he snaps.

I feel lonely without him around, but there's more peace in the house. Kevin and I do more together. Kevin's taken a yen to baking and has become expert at chocolate chip cookies—a hobby I encourage—while I fix up his room. Months ago, John carved him a Viking headboard, so I'm now building the child a bed to attach it to, painting his room and building him a desk. Meanwhile, every second weekend Kevin goes to his mum's, and John and I retreat further to our separate spaces.

One of the things John likes to do is lie on the bed in our screaming orange bedroom and listen to jazz on the portable radio. Jazz is incomprehensible to me: where's the tune? One Saturday afternoon, when there's a worse squawking than usual coming out of the bedroom, I ask, "What is that?" John lifts his head as if from a trance.

"John Coltrane. Haven't you ever felt like that?"

So that's how you're supposed to listen to jazz! It's metaphor again—using one thing to express another. I just needed a door. But moments like this between us are rare. Every day now, John leaves the house immediately after breakfast. If he doesn't stay at school late to study, he stays up late at home. Even when his body is sitting at the supper table, some vital part of him is buried in a law text with moot and appellant and *ad idem*. He's in a country as new to him as construction was to me, learning a language that excludes Kevin and me.

"A lot of marriages break up in law school," he announces cheerily one night at supper.

As tension mounts at home, some days I feel I must break. If I ask, "Are you in a bad mood?" John snaps "No!" and walks away. I wish for a body stern as steel. John yearns only to sail his newly finished boat, but after one trip with him I refuse to go again because it's too scary; he doesn't know how to sail. The only thing we share now is lacrosse. Kevin is entranced by the game, and though I never do get too clear on the rules of play, I'm happy to sit beside John, sharing the enthusiasm of real parents inside the airy echoes of an arena, hiding my eyes every time another huge lacrosse player races toward my stepson, the goalie, aiming a ball as hard as a baseball at his thinly helmeted head.

Every evening, I write in my journal. Writing seems to relieve the pressure, help me figure out what happened that day. When a local women's magazine asks me to write about my construction experience, I write about the first day I stepped onto a construction site, on the island. It's only in writing it that I realize I was learning a certain grace. "The Zen of Construction" is my first publication. The more tired I am after work, the more condensed these journal entries become. The shorter lines look almost like poetry, and I begin to write poems deliberately, condensing even more. I don't show anyone; I write for myself.

The other thing I like to do after work, especially on Friday nights, is swim. The long, slow strokes of crawl and breaststroke at the community pool are deeply relaxing, especially when I finish with a soak in the hot tub. Without my glasses, the world is reduced to rough shapes and blobs of colour, and one night as I lie in the hot tub after a swim, I notice that the curves of the heavier women are far more interesting, more beautiful, than the stick figures of skinny women like me. When I get home, I write down the line that's been making slow circles in my head: "Some hips are made for bearing children." Then another, and another, until there are no more lines in my head. When I read it over, this, I know, is a poem. I title it, "These Hips."

> Some hips are made for bearing
> children, built like stools
> square and easy, right
> for the passage of birth.
>
> Others are built like mine.
> A child's head might never pass
> but load me up with two-by-fours
> and watch me
> bear.

When the men carry sacks of concrete
they hold them high, like boys.
I bear mine low, like a girl
on small, strong hips
built for the birth
of buildings.

how is a woman like a hammer?

It's at one of our many bad-news union meetings that I meet Chryse, the second woman to join our Vancouver local. She's slim, fine-boned, quick and light in her movements, with dark blond hair pulled back in a ponytail. And she's already a journeywoman.

When Chryse suggests we do odd jobs together, I look at her stupidly. I barely know what an odd job is, let alone how to do one, but my UI cheques are running out, and at least odd jobs aren't like working for a big non-union company. Chryse got her papers in New York. Whenever there was no union work there, she tells me, she and her husband did renovations. When the union agrees that any hours I work with Chryse can be applied to the two thousand I need to finish my apprenticeship, I tell her I'm game. Now, how do we get an odd job?

"Leave it with me."

Sure enough, a few nights later she calls to say someone wants some renovation work done, and we need to submit a bid.

When we meet at the house, I'm dazzled by her confidence. She takes a pad of paper and a pencil out of one pocket, a tape measure from the other, hands me the paper and pencil and starts measuring, calling out numbers. A few days later, the owner says yes. We have a job.

The owner, Julius, is a labour activist and filmmaker who thinks it's great we're union carpenters, and women. He wants a new wall with a door in it on his main floor, and upstairs, a new study.

On day one, I pick up the materials on Chryse's list because I'm the one with the truck. We unload it together. This time it isn't a question of blindly doing what I'm told. This time I'm a "partner"—sort of. I not only have the truck, I have experience with concrete and framing, while most of Chryse's work in NY was fine finish and drywall.

Our days take on a routine. We arrive at work at 7:45, perch on the sawhorses, open our thermoses and plan the day. At eight o'clock we lift our tool belts off the nails where we hung them the night before and get to work.

It's vastly different to work with a woman. For one thing, we talk. Chryse works out loud, thinking and explaining as she goes. With Chryse, being a woman is normal. I don't have to control my questions, don't have to remember not to giggle, not to be provocative, not to say or do anything "feminine" that would draw attention. I can use all my energies, both the male and female parts of me. The first time we get distracted by how interesting our conversation is, Chryse says, "We're slowing down. Work like a carpenter." And that night we phone each other to talk—like women—about the personal stuff.

—

A week or two into the job, we've demolished the old office and framed in and drywalled a new one. Now, as Chryse tapes and plasters upstairs, I frame in the new wall on the main floor. I snap a chalk line where Julius said he wanted his wall. It's an idea older than the pyramids that the shortest distance between two points is a straight line. I cut two-by-fours to length until a double row of plates lie along the chalk line, hook my tape over the end of one plate, make the adjustment for the first stud, and mark sixteen-inch centres, walking as I mark. In the absence of blueprints, I pick what I think is a good location for the door, in line with the front hallway so there'll be light and openness in the new room. I mark both plates, move the top one until it's roughly eight feet from the bottom one, fill in and nail the studs, then call Chryse.

A male apprentice has to prove he can't do this job, but a woman doesn't get the benefit of the doubt. With the men, I've had to prove I can do it, every time, and I hadn't realized how exhausting that was. Until now. Chryse just assumes I can do it.

With one of us at each end, I give the signal, and the wall comes up. Chryse holds it steady as I knock the bottom plate to the blue line and nail it, then plumb both ends and nail them to the walls. I can feel by the bounce of my hammer that the nails are going into nothing more solid than the lathe under this old plaster, but that's okay. I'll nail the top plate to the ceiling joists later. When we step back to admire, Chryse says, "I like where you put the door."

When that job is finished, Judy, from Women in Trades, asks us to help her do a reno putting a foundation under an old house. While we're at it, the owners want a fence around the property, and a patio. Since neither Chryse nor Judy has done much concrete work, I'm in charge— my first job as a forewoman.

To lift the house, we all agree we need the experts. Zebiak Brothers is the sign I've seen at house raisings around town, and Mr. Zebiak turns out to be not only a pro but a sympathetic one. A couple of times he even gets his crew to do extras for us. When I thank him, he leans on the fence, looks at his feet and says, "I've got a daughter who's a terrible libber."

Once the house has been raised, we get to work. It rains every day. Chryse tackles her first set of footings while Judy and I hunker under the house in a mud bath with two feet of headroom, working to replace rotten joists and dig out for the new foundation.

Chryse's tickled with her first set of forms. "They're so cute!" she keeps saying. "I feel like a real construction worker."

But we aren't very construction worker–like. When Judy uncovers a mixing-bowl-sized seething bed of tiny red worms, we scream. Then laugh. And after crawling all day in the mud, Judy emerges singing, "I'm a mole! I'm a mole!" and does the shimmy on the lawn.

I politely consult everyone for all my decisions, including the clients, Jim and Martha, who every morning feed us healthy muffins and good strong coffee. "We'll do so-and-so today," I say, then, "Okay?" Everyone feels obliged to comment, and some days these consultations take up to an hour. An unpaid hour. Finally, I remind myself to think like a foreman and stop asking rhetorical questions.

It's nerve-racking, calculating the amount of concrete we'll need. Over and over it comes out to seven and one-half cubic yards, but I know even experienced contractors are anxious when they give the cast-in-stone final figure to the concrete company. Too much, and you waste money. Too little, and you don't have enough for the pour. In the end, I order seven and three-quarter cubic yards and persuade the clients to agree to let us hire a pumper truck and a concrete finisher, the latter so the patio will look good.

We schedule the pour for noon so our clients—who insist—can be there to help. By one o'clock, when the truck grinds up the alleyway, I've taken people through a dry run of what will happen and given everyone a job. Vern, the finisher, arrives just before the pumper truck. The concrete driver is clearly shocked by our unseemly crew: four overly eager women, a black tradesman and a scrawny white guy. He keeps addressing his questions to Jim, but Jim doesn't know the answers. So he tries Vern. Finally, in defeat, the driver talks to me.

The job goes like magic. Our forms don't so much as twitch when the concrete flows in like grey thread from the belly of a silver spider. Vern's done concrete for thirty years and has some good suggestions, so we jiggle tasks and by the end of the day the foundation is poured, the patio shines and the fence posts stand straight as soldiers in their little concrete beds. Even the truck driver gets into the spirit; when we go fifteen minutes over his waiting time, he doesn't charge us, and there's less than a quarter yard of concrete left in the truck. But the highlight of the day is when I mention to Vern that the pour seemed to go well.

"That's because you had such good forms," he says. Later, he'll tell Jim they were the best forms he's ever seen. He wonders who trained me. But Vern is no slouch, either. His suggestions were exactly right, his hand certain, and the final job a beauty. More, he's fun to work with. From now on, every time I pour concrete, Vern will be there.

I go home that night feeling splendidly androgynous. I am ferocious lion, sweetest lamb and sensuous snake—as I choose. I feel 100 percent alive.

Only when the job is over do we find out our clients got six bids for the job and ours was the lowest.

—

Chryse and I bid on two more jobs and win neither, but I'm not the daughter of a marketing man for nothing. I get the bright idea we should advertise.

Chryse isn't so sure. "You think people would hire a woman carpenter out of the newspaper?"

When we lose our third bid, we each chip in twenty dollars to place an ad in the local paper, being deliberately ambiguous about the "woman" part. "Carpenter and crew available for additions and renovations," the ad reads. Okay, we exaggerate "the crew" part, too. We get one bite, from a guy who asks to speak to the carpenter. When I say "Speaking," he says, "No, I want the carpenter who put the ad in the paper," and it goes downhill from there.

Chryse laughs and says it only pays to advertise if you're selling a white, male product. We're both thinking of Carlyal, who recently joined Women in Trades. Carlyal is six feet tall, black, with a curly head of hair that stands out in a dramatic Afro. She loves bricks, and to look at her, you'd think she's a natural, but she's told us that most of the men on her bricklaying jobs wouldn't help her. Labourers were slow to bring her mud, and when one of them "accidentally" dropped a two-by-four on her head, she gave it up and switched to tile-setting.

"Though I don't know why they think women can lay tile and not brick," she says drily. "What do they think, we carry all those tiles in one at a time?"

When I drop into Dispatch at the union, I learn that though our local has the lowest unemployment rate in the province, we're still 51 percent unemployed. Jobs are now being split into three-week chunks to help spread the work. We lost the legal battle at Pennyfarthing, and non-union companies are winning contracts for commercial and industrial jobs. Under another new government policy, public bids now automatically go to the lowest bidder. Even I know how to keep a bid low: poorer materials, lower pay and more speed in building. Which means more risk for the men and women building it.

—

Chryse and I win our fifth bid—a basement renovation that includes lots of concrete. We get the job through an architect named Shirley who's also just getting started. We hire Peter, a union apprentice smaller in stature than either of us, to do what Chryse calls "grunt labour." Peter announces himself on the first morning with "Here's your brute!" I show him the concrete slab we need removed to make room for the new basement extension, then go off to do interior demolition while Chryse makes phone calls upstairs.

A contractor is someone who expects the unexpected. I'm not a contractor yet.

One hour later the concrete's all gone and what's left is a short, shiplapped wall resting on dirt.

"We better see what's in this wall," I say a little anxiously. But we've barely begun pulling off shiplap when Chryse comes screaming downstairs.

"Cracks are opening everywhere!"

The beam carrying the kitchen above had been supported by the wall whose bottom six inches—we can now see—are completely rotted away. The only thing holding up the kitchen above us is the shiplap Peter and I are now removing. My father once told me, "To assume makes an ASS out of U and ME."

That morning I'd rented pole jacks for when we'd need them later, in the structural stage. Clearly the structural stage has arrived sooner than expected. We whip the jacks under the bearing beam, Chryse goes upstairs with the level and Peter and I adjust jacks until she yells that the kitchen is level again. And that is how I catch the renovation bug.

—

The estimate for the heating work is six hundred dollars. "For three vents?" the customer asks in disbelief. So I consult the Yellow Pages and call a second tradesman. His name is Mickey, and when he arrives, he seems an unlikely prospect. He rarely makes eye contact and ignores us completely except to probe in detail about precisely how the client will use this space. A few scribbles on a scrap of paper and he says, "Two hundred dollars."

He may be odd, but I like him, and it quickly becomes clear he has a genius for heat. He's also much cheaper than our last bidder. When Mickey's done, his vents are beautiful, and when we tell him what the first bidder wanted, he laughs.

"You would have been paying him $100 an hour."

Chryse and I start paying Workers' Compensation coverage and looking for a company name: KayCee Construction, for Kate and Chryse? Wild Women Construction? Maids in Canada? My personal favourite is Tough Tulips Construction, but everyone laughs too hard. Finally we settle on Sisters Construction because we're both in the union. We have infrastructure too. Sisters buys a company wheelbarrow, and when parts start falling off my truck, I get my first bank loan so I can buy a ten-year-old Datsun pickup that—unlike the first truck—stops when you put on the brakes. I lie through my teeth and tell the loans officer we make $1,600 a month and have been in business for a year. She doesn't ask for proof. One day it will be true.

The only thing that doesn't feel good these days is that I keep being mistaken for a man. In the food store, a woman brushing by says, "Excuse me, sir," takes a second look and apologizes profusely. Even Vern, the cement finisher, when he sees me carrying his gear to the car, thinks I'm a guy. Am I losing my femininity? What is "feminine," anyway?

Our cost estimates for the basement job are good but we're way behind in our time estimate, and the client has promised relatives they can stay in the new basement in three weeks. Soon we're working six days a week from 7:00 to 5:30, paying ourselves two hundred dollars a week each, until the advance runs out.

In the past, Craig—Chryse's husband—made all the structural decisions when he and Chryse worked together, but he and I increasingly cross swords. One day after he stalks out, angry at me again, Chryse and I take off our tool belts and talk. She says I offend people when I say critical things without looking them in the eye. I tell her that at those times, I'm not thinking about feelings at all, that the men I've worked with didn't care about feelings, either. But now when I think about it, the best foremen were subtly paying attention to the emotional parts too: who worked best with whom, who didn't mind doing which work... So she's right. Supervision, too, is a learned skill.

The kitchen cabinets go in, stove and refrigerator. We pay a cleaner to clear up, and by late in the afternoon of June 15, the new suite is finished and the clients are happy. We've done it, to the day. And I vow never to do it this way again. Since we'd signed the contract, the clients had made numerous additions, and it's only now that I wonder: why hadn't we explained that the new requests would take more time? Why hadn't I just said no?

Activists in the province are in an uproar. The Social Credit government under Bill Bennett has taken advantage of their recent re-election by passing—along with a new budget—twenty-six bills that aim to abolish almost everything I care about, including many of the ways women first accessed trades, like the Human Rights Branch. Union and community groups are forming a coalition to protest, and the Vancouver Status of Women calls a meeting of every women's group they can get hold of at short notice to discuss women's input. We all know how important this coming together is. We're excited about fighting back and about the extraordinary fact that the unions want to fight with us. That's a first, and it gives the women's and community groups a greater sense of power and possibility. There's fear, too. We all know the kind of resistance there is to women's issues in the unions. But for now, we have important decisions to make; for example, should each separate women's group ask for representation in the new coalition? Women sit forward in their seats, hands raised, clamouring to speak. This time, we decide, women should speak with one voice. We call ourselves Women Against the Budget (WAB).

In mid-July, the Lower Mainland Anti-Budget Coalition meets for the first time. We're in one of the biggest union halls and still it's jammed. The energy is electric. It's soon clear that the community groups are as worried as we women are that the trade unionists, with their overwhelming organizing experience, will try to dominate. On every issue, there are long lineups of people who want to speak: feminists, trade unionists, members of church and community groups, liberals, communists, conservatives, socialists of every stripe, and many who aren't affiliated with any group who simply don't like what this government has done. There's cheering and applause and the odd boo that's quickly shushed. People remind each other: we're working together now. This moment is historic.

After picking a chair (George Hewison, president of the Fishermen's Union) and settling on process, we move on to the vital question—what will we actually do? After more discussion, we agree to hold a rally on July 23. Art Kube, the head of the BC Federation of Labour, dubs our loose organization the Solidarity Coalition, after the Polish workers who've recently won democracy, and the name sticks. Sometimes, once things get moving, we're also called Operation Solidarity.

The next few weeks are filled with endless telephone calls leading to endless meetings and some actions. At one point, Women Against the Budget puts on a soup kitchen. If the government won't feed its needy, we will.

The Coalition's demands are simple: withdraw the bills! There are rallies all over the province. In Vancouver, where organizers had hoped for ten thousand, the unions count twenty-five thousand. In Victoria, thousands more take the afternoon "off," and twenty thousand turn up for the biggest rally in the city's history. We all watch nightly news reports with our breath held. Is it possible we're actually pulling this off?

The main rally in Vancouver is planned for August 11. When we hear that it's booked for Empire Stadium—a place reserved for major sports events—some of us figure we've gone too far. How can we possibly fill all that space? It's a weekday afternoon and most people will have to leave work, lose pay. That's especially hard for public servants.

Half an hour before the rally, I walk into the stadium with the other carpenters and catch my breath. Not just the stands but also the grounds are packed with people cheering and waving banners that show their affiliation. I've never felt so proud, so strong. Then the gates at the far end of the field open, there's a blast of music, and in marches the firefighters' band. An essential service taking the day off, breaking the law to be with us! Most of us have tears in our eyes as forty thousand people spontaneously rise to their feet, cheering wildly as the band marches in, followed by dozens—no, hundreds—of firefighters, then bus drivers, all in uniform, then the Canadian Union of Public Employees. Solidarity indeed.

—

At the end of June, I'd received a letter that began, "Dear Sir and Brother," telling me I'd completed the hours required for my apprenticeship. I am now officially a "journeyman" carpenter, and two weeks after the rally, we celebrate. My friend Claire designs a cartoon invitation that ends, "Bring family and come watch us roast a turkey (Bill Bennett excluded)." The "ex" referring to our province's unpopular premier has been crossed out and replaced with "in." The tradeswomen come, and my union brothers, neighbours, some of my new friends from Solidarity, even clients—altogether some sixty people gather in our backyard to help celebrate. John and I make a gallon of bean salad and bake spinach pies and a brandy pumpkin cheesecake, and barbeque a turkey and two salmon. The guests bring salads, desserts—and presents. I hadn't expected those. The most moving is a set of fine finishing chisels from the WIT women, wrapped in a leather case made from Alice's—our motorcycle mechanic's—ex-riding jacket. My parents' one-and-a-half-inch chisel completes the set. I am now a fully qualified Red Seal Construction Carpenter, with a beautiful set of chisels.

—

After the basement renovation, Chryse and I have a few small jobs lined up, but when she gets called for her three weeks of union work, she takes it, of course, and is glad, not just for the money.

"I'm not in love," she tells me on the phone the following weekend, "but I have a mad crush on this guy I work with. We're such good partners. He's so good at what he does." Her pleasure is not just in the work—she's deeply missed the fine finish she's doing now—but in doing it together. "And I feel more feminine when I'm the only woman there."

It's one of the ways we differ.

—

In an attempt to ease the tension between us, John and I plan a vacation with Kevin before the two of them go back to school in September. We're also celebrating a new agreement that Kevin will live with us for the next five years. We'll go camping and take our nephew, Eric, who's ten. But on the morning we're due to leave, John takes Kevin out for a quick sail. When they're not back by 1:00, I phone the sailing club. Nobody answers. At two, the phone rings. It's Kevin.

"We're in the hospital. My dad can't move his legs."

In trying to dock the boat, John damaged his back and is immobilized with a herniated disc. He comes home, but for seven days, even with heavy painkillers, he can't sit up. I move to the couch so my movements in bed at night don't cause him pain, but still he wakes us with his moaning.

I keep Kevin and Eric distracted with shopping, cooking, movies and canning. The boys and I buy boxes of fruit and vegetables, and for days we slice and peel and measure and stir, standing over bubbling pots, canning quart after quart of pears, peaches, tomatoes, beets, beans, dill pickles and jam. Building walls of food seems one way to protect my family.

On the first day John sits up, his mother tells us her doctor has discovered a malignant lump; she's going in immediately for surgery. John's still bedridden, so I take a day off work to register him for his second year of law school, and every morning, I drive Kevin to his new high school on the west side of town. After work—a renovation on the east side—I visit John's mum in a North Vancouver hospital before coming home to cook supper, help Kevin with homework, make the necessary phone calls for work the next day, see to John's needs and fall into bed. At work I get stung by a hornet and my right arm and hand swell up, but I can still hammer. I operate out of my steel core and do what has to be done. Then Mr. Moon, the cat, dies.

With Ruby, the dog, beside us in the garden, Kevin and I have a lovely ceremony in which we talk about all the good times we've had with Mr. Moon. We officially forgive him for eating Kevin's gerbil before we bury our beautiful white cat near the tomatoes. It's a good excuse to cry.

Kevin is thirteen now, almost six feet tall, thrilled to be picked for his school's basketball team, fun to be with. One day as I'm making supper, he comes up close beside me and whispers in my ear. I'm paying attention to the unusually worried look on his face so at first I don't catch it, and he has to repeat, "I have a pubic hair."

Not sure what a real parent would say at this important moment, I tell him the only thing I can think of. "Congratulations. You're becoming a man!"

As I watch him head back to his room, apparently consoled, I'm so honoured to be entrusted with this important information, so glad to be this lovely young man's friend. These days he calls me his co-mum. I call him my chosen child.

By October, John's back in school, and I have no more work lined up, so when Dad offers, I let him buy me a plane ticket home for Thanksgiving. I celebrate by getting sick, and as I recover, it's deeply healing to hang around all day with my mum and dad, talking and eating and drinking tea. Even with Dad home, newly retired, the house is strangely peaceful. He still drinks, but mostly wine, and not so much. When Mum mentions that she wants all the door handles changed and I offer to do it, it's Dad who drives me to the hardware store, and we end up doing the job together. As a family, we watch TV and look at family photos, and occasionally, Dad lets slip stories about the war, how at age twenty when he was already a flight instructor, brought in from Britain for the British Commonwealth Air Training Plan in Alberta, he was training other pilots even younger. He talks about death, about how often students crashed after the mere hours of instruction they were given before being sent back to Europe to fly bombers and jets. Reluctantly, I begin to understand why he might have sought solace at the bar.

—

The carpenters in my local have elected me as a delegate to the annual Carpenters' Convention, and two days after I return from Montreal, I leave for the convention in a nearby town, where I'm plunged back into the intensity of Solidarity. The government has ignored our protests and tension has been growing, until now there's talk of a general strike. No one wants it—there's too little work as is—but how else can we make our point?

I faithfully attend every session at the convention, and speak up. On the first day, I get invited to something called a "caucus," in which the Leftists get together to plan strategy. But I don't agree with everything the caucus decides, so sometimes in convention I don't vote with the rest of them. After the first day, Lorne, the secretary-treasurer of our Provincial Council, takes me aside to explain why it's important that all of us in the caucus vote together. He's a big man who speaks forcefully and I'm grateful for the practice I've had in standing up to my dad.

"If you belong to the caucus," Lorne repeats, "you agree to vote the way the caucus decides."

"But how can I vote against my conscience?" I gather this is not acceptable Leftist behaviour, but I don't stop. How can I? Too often, doing what powerful, articulate men have told me is the right thing has left me feeling very wrong.

The decision is made in Solidarity to start an escalating series of strikes. One of the main targets of the government's legislation was the BC Government Employees' Union, and now the government threatens that at midnight on October 31, the minute their contract expires, 1,600 BCGEU members will be fired. The union promptly counters by going on strike. A lot of us are surprised when the next group picked by the Solidarity leaders to strike are the teachers. It doesn't make sense. The BCGEU are a militant union, but the teachers won't even call themselves "union"—they're a federation. There are more meetings, more phone calls, and much muttering that our leaders are getting cold feet. If the teachers don't walk, it will be the end of Solidarity.

The day chosen for the teachers' walkout is November 8, and tension mounts. Will they go out, or won't they? At two that morning, I get a phone call.

"Teachers are walking!" an urgent male voice says. "They won't go to work, but they won't picket, either. If we don't have pickets at every school by 8:00 a.m., it's over."

By 7:30 a.m. I'm standing in front of an East Vancouver elementary school with people from a dozen different unions. Someone hands me an On Strike sign and brochures to pass out. It's a bit scary because no one knows what the individual teachers will do. What if they come to boo, or to cross our lines? And they do come—bringing coffee and doughnuts. Some of them cry as they stand beside us to explain to each wide-eyed kid who shows up, what a union is, and why they are doing this. Sometimes, listening, the rest of us cry, too.

In Kelowna a short time later, there's a late-night meeting between Jack Munro, representing the union leaders, and Premier Bill Bennett. And suddenly it's over. A deal has been struck. For many of us, the Kelowna Accord represents our worst fear—a sell-out of community groups and women by some of the union leadership. Though no trade unionist I know is happy with the deal—or the way it was semi-secretly arranged—we go back to things as before, only a little more bitter.

—

In November I'm dispatched to a three-week union job, building cross-heads for the new Vancouver Rapid Transit system. I'm so excited about making real money that I don't have time to be scared. At 7:30 the first morning I walk into a shack filled with twenty carpenters. After a moment of dead silence—I'm getting used to this—someone says, "Ho-ho-ho," and the chatter picks up again.

This job feels disorganized. It's an especially cold and rainy winter, but the company hasn't bothered to put heat in the shack, and they don't have a Johnny-on-the-Spot. On day one, the foreman lends me a truck so I can drive to the nearest gas station when I need to pee, but after I try it once, I tell him this is silly.

"I'll go behind the columns, like the men." And honestly, I don't mind, but the next morning we have a Johnny-on-the-Spot.

At first they put me in the yard building forms with an old guy named John. It's easy work, so I figure they're checking me out. At the end of the first day, John says without hostility, "You're doing as good as the last guy we had here." But the next day, after he's had a chance to think about it, or has perhaps told his wife he's working alongside a woman, things don't go nearly as well: suddenly John needs other carpenters to help him, and he makes jokes I don't get, but don't like.

The next day they send me up top, to work on the column heads.

In this wet weather, after the steel forms have been bolted and the concrete poured, the metal nuts are often badly rusted when we come to strip them. It's tricky, working twenty and thirty feet off the ground, to get the leverage to pry them off. That is, until I find a can of WD-40 in the tool shed. At first the guys scoff, but when they see how fast I can crank off the rusty bits, everyone wants a can of WD-40 of his own.

Still, not much seems to be getting done until I'm assigned to a new foreman, building bulkheads. I like how this man works, and eventually he waves me over and the two of us work as partners, finishing a whole bulkhead while four carpenters struggle with a second. It's fun at first, but as the days pass I catch him looking my way as if to ask, "How do I relate to you?" Then comes a period of "accidental" touching, and after that, some swearing at the boards, more aggressive than it needs to be, as if he's nervous, as if he's saying, "I am a man!"

As if I've questioned it.

After three weeks, I'm laid off, as expected, and a couple of weeks later, Shirley, the architect, phones to offer Chryse and me a chance to bid on a kitchen renovation.

"If your estimate's in the ballpark," she promises, "this guy will go with it."

Our rough bid is accepted but we've agreed it's my turn to do the detailed estimate. Monday morning, I buy a shiny yellow spiral notebook and begin to fill page after page with lists of time, money, materials. By Tuesday afternoon the pages crackle like parchment. When it's finished, the bottom line is a bit alarming, so I double-check everything—still $8,500.

"That's what it costs," Chryse says.

I know I'm fearfully ignorant as a contractor, especially of anything to do with electricity or plumbing. I try partly to fake it, partly to fly by the seat of my pants, partly to ask what I hope are intelligent questions. My initiation comes when I ask D., who does both electrical and plumbing, to give us an estimate. He starts by not being able to find the electrical box. Bad sign. His estimate, when it comes, seems very high, and when I check some of his prices at the retailers, they're also high. I get a bad feeling when his explanations leave me more confused than ever. (I've never heard of contractors working by intuition, but this one is.) I pick up the Yellow Pages and, under Electrical Contractors, look for a small ad (I figure that shows low overhead), for a European name (because Europeans are the best trained), and for someone who's been in the business a long time (which suggests quality).

When Ivan comes to look at the job, I shake hands with a small, quiet Dane to whom I take an instant liking. His bid is a lot lower than D.'s, and when I tell D. he's lost the electrical bid, he says gruffly, "Then you better get yourself a new plumber too," and hangs up. So in mid-December of the coldest winter in forty years, I start to phone plumbers. But they're all out on emergency calls, and it's only when a frozen pipe bursts and the basement in which we've just stacked our drywall is floating in three inches of water that I manage to lure Gino—on Ivan's recommendation—to fix it. For the whole forty-five minutes that Gino works on the broken pipe (someone else's mistake, thank goodness), I sweet-talk, cajole, chat, almost beg him to take our job.

Finally he mutters, "I'll look," and grumbles his way upstairs while I clean up the mess. He emerges smiling. Chryse has worked her charm, but I'm on eggshells until he actually says, "I'll start tomorrow." When Gino first looked at this job, he'd clearly been nervous about us being women. "Never seen this before..." but by the end of his second day with us, he announces, "it's marvelous!" that Chryse and I do this work.

This is how we put together a crew, a team of people we like that includes Ivan and his son Per as electricians, Gino for plumbing, Mickey for heat, Roy for glass, Prebin for drywall, Carlyal for tile and Shirley as architect. We officially register our name, and on December 16, I make the first deposit into the brand new Sisters Construction bank account.

Contracting, Chryse says, is about decision making; "Just make a decision," she tells me, "any decision, then live with it. They aren't mistakes if you fix them, build around them differently than you first planned."

Still, I can't stop worrying that others will have to live with (and pay for) any bad decision I make. It paralyzes me and my delays slow everybody else. And the decisions are unending. Where does the stove vent go, the electrical apprentice wants to know, and I'm getting good at answering, "Standard height," but he doesn't know the standard height and neither do I. Chryse leaps into the gap and gives a number that— only I know—allows for the two-inch mistake she made earlier in hanging the cupboards. She whips a mark onto the wall.

"Here," she says, with what looks like total confidence. "It should be a little lower than usual to get all the cooking fumes." And the electrician happily proceeds to cut a hole in the wall.

I compensate with obsession over detail. One day Chryse overhears me on the telephone with the linoleum contractor ("But if the floor needs patching? If your floor layer doesn't phone when he says he will ...?").

"Leave him alone," she says. "Trust him." If you pick good tradespeople—which we have—and clear the way for them to do their job— which we do—then they'll do it just fine.

My training has prepared me to handle all the parts of contracting; the challenge is to put the parts together and keep all the balls in the air at once. When we talk about contracting at Women in Trades, Judy, the carpenter we worked with on the first concrete job, says her experience as a mother of three was the best training she could have had for contracting: multi-tasking? No problem.

I'm also beginning to understand that what I always thought of as the men's bullshit isn't merely—or not only—lying. Ask them if they can do something, they invariably say, "No problem!" whether they can or not. Now I, too, increasingly say "Yes, I can," to jobs I haven't actually done before. I'm beginning to have the confidence to know I can figure it out.

I promised our client we'd be finished by mid-February, and on Friday the thirteenth there's a bustle of trades and chaos of a joyful order until, at 7:30 that night, we shoo everyone to one end of the brand new kitchen and turn off the trouble lights and, in the dark, switch on the bright yellow lamp hanging in the new nook and everyone says, "Aah!" because yes, it is beautiful.

⚊

Before we start the next job, Chryse and I discuss our company. She doesn't like small renovations—the constant change of pace, having to deal with sub-trades, clients and architect, the structural work—all the things I relish. She likes fine finish, the routine of big jobs, lots of guys, the politics. Our styles of working, too, are different. I'm the battleship who anticipates conflict and ploughs straight through the front lines. She's the snappy little powerboat who zips around behind, picking up survivors. "You're more into contracting, the business side of renovations," she says, and it's true. We decide that, until she's called again for union work, I'll continue doing the contracting and Chryse will be chief carpenter.

When I worry about doing non-union work, Bill of the Provincial Council encourages us. "People have to eat," he says. Union work is mostly industrial and commercial, so the union has no objection to their carpenters doing residential jobs. Besides, there's talk of one day using the union's pension fund to finance its own construction company. "So go for it."

Things are getting worse for the union all the time. Two members are rumoured to have committed suicide, and we hear that union carpenters are taking non-union jobs for as little as $4.50 an hour even after providing their own power tools and cords. Chryse and I made $6.66 an hour on the kitchen job, so I guess we're doing okay.

In late January, my unemployment insurance runs out. I've covered everyone else's bills for our latest job but haven't paid myself since the end of December. We seem to have no problem anymore getting jobs—mostly through word of mouth—and I'm working six days a week, ten hours a day. Even when I'm not working, the job never leaves me. John's and my last lovemaking ended abruptly when I wondered out loud how I was going to waterproof the join around a toilet. In fact, between John's focus on law school and mine on construction, our relationship is floundering badly. He agrees to go to counselling with me, but apart from making us more civil with each other, it doesn't change a lot.

One afternoon in early spring, when we're fixing supper, we have our worst fight ever. For the thousandth time, he doesn't seem to hear what I'm saying. Nothing gets through to him. Not calm reason, not yelling at the top of my voice—nothing—and in a moment of pure rage, I hit. I aim at his face.

He easily blocks my arm. His face is frozen inches from mine, and we're locked together for what seems like minutes but is surely only seconds. His eyes are deep brown and strangely clear. As surely as if we've spoken it out loud, I know we're going to split up.

13

doing it right

Two days before Valentine's Day, John and I agree to end our six-year relationship. The heaviness I've been feeling disappears overnight. We plan to stay in touch, and Kevin will visit me once a week. I move out—to housesit for a friend—while John finds an apartment for himself and Kevin, and I look for a roommate to help me pay rent when I move back in later.

Now that the pressure's off, we can finally enjoy each other. The three months it takes John to find a new place are almost a honeymoon, but we're both clear this is a final fling. When he and Kevin move out and my new roommate, a grad student from the university, moves in, I spend a week redecorating what is now my room. With it comes the full impact of separation. At night I dream of houses with no foundations sliding down mountains. In the day, the pain comes in spasms, a physical grief that slashes me from head to foot so that I have to hold on to something solid until it passes. When I ache too badly, I phone John and am lifted by the joy in his voice at hearing from me. Small talk feels big enough. Sometimes we even sleep together.

I start swimming regularly, and lifting weights again. For the first time in my life, I wear a bit of red, and dangly earrings. I miss John, but not for a second do I think of going back. Instead, I go to a therapist to try to make sense of this difficult change that feels so right.

I bid on, and get, three small jobs, so I have enough money for the next month's rent, but my income is minimal, so when I get a letter inviting me to speak at an international women's conference in Groningen, the Netherlands, about women in non-traditional jobs in Canada, my Women in Trades group decides they're going to get me there. They rent a local coffee house, put up posters and hold a benefit auction at which the tradeswomen auction off their skills: a motorcycle tune-up from Alice, and another for a bicycle from Joan; computer help from Janet; Judy's carpentry skills; and the pièce de résistance, a gourmet dinner for four cooked in your own kitchen by Julie, the woman in charge of women's issues at the province's Ministry of Labour, complete with a bottle of the same expensive wine over which there's recently been a government expense-account scandal.

The room is packed, buzzing, and I sit at the back in a state of wonder at having such friends. In these days of high unemployment, they manage to raise $1,500 in a single night. I'm going to the Netherlands, and to top it off, Janet, the avionics mechanic who works for CP Air and gets travel passes, will come with me.

At the conference, I'm the only woman who actually works on the tools. The rest—academics and government employees from the Netherlands, the United States, Denmark, West Germany, Italy and Norway—report a strikingly similar list of problems to ours in Canada. It's depressing to know that sexism has such a global face, but when Janet and I travel after, it's exhilarating to meet European sisters in trades. If we ever doubted it, now we know for certain that getting women into the trades is neither a small nor a local issue; we're taking on human history. It will take time.

⟶

In June, I get dispatched to my first long union job in two years. The company is building a high-rise in downtown Vancouver, and it's a relief, when I walk into the carpenter's shack, to recognize two of the men. When they say hi, it makes it easier for the others to accept me, too.

Now that I'm journeyed, the foreman gives me my first apprentice and puts us to work building scaffolding. The apprentice stays on the ground while I go up top, fifteen feet, to haul up the flat heads as he ties them on below. A flat head is a sort of metal cup that, when you set it into each corner of the scaffold, holds the beams that support the next floor.

The crane operator must be having a slow day because suddenly I hear his honk and look up to find the crane's hook at my shoulder. We work out a system: the apprentice ties two of the heavy flat heads to the crane's hook while I place the joining cones that will tie them to the scaffold. Then I hold my arm high so the crane operator can see it, drop my wrist and twirl my index finger as if stirring something as the operator lowers each flat head exactly into place. When the foreman sees how smoothly it's going, he laughs. But it starts again—it always starts again—when the crew realizes the implications of a woman doing this work. I'm doing it; I'm even good.

"What will you women want next?" the labourer foreman snaps next time he walks by.

At first I wasn't too sure about Tom, the foreman, but as the days go by he doesn't ride my back or ignore me or uselessly pace when the work isn't going fast enough; he actually helps. "What do you need?" is his line, and suddenly there's the saw blade or two-by-six I needed. So I'm not sure why I feel cautious around him.

One day in the shack, someone mentions the name of the supervisor on my first union job—the first high-rise. Before I know it, I'm telling them what that supervisor said when I left, that I didn't really think this was women's work, did I?

"So I said, damned right it's women's work," I wind up. I've blown it now, letting my feminist politics out of the bag. But one guy adds, "And get a decent wage while you're at it!" Someone else gives a long speech about his sister, who's head teller in a bank and gets paid less than the male teller she trained, so now he's her boss. I've never heard guys in a construction shack talk like this.

When we're back at work, though, the job steward says, "But you have to admit—a crew of women couldn't get this building up." This is the guy who's supposed to be my special ally? When I just stare at him, he adds, "In as short a time, I mean. It takes muscle."

"Women have muscle, and leverage."

"But it takes a lot of muscle."

"And brains."

"Mostly muscle."

Then the ironworker foreman jumps in. "What about good management?" Later, the same ironworker, working a few feet away from me, says quietly, "I've never worked near a woman swinging a hammer before."

"It's like anyone else with a hammer," I tell him. "Keep your hard hat on."

I'm careful, knowing how important it is that their questions and doubts be addressed—and how rare it is for them, how precious to me, when they say out loud what's really on their minds. I never know how they'll take my answers, but as these exchanges keep happening I'm humbled at the implications. These guys are with me on the cutting edge of change. I'm touched by their grace.

These days I go home exhilarated. The sun shines, the breeze blows, and we're into the thrill of height now, two stories above ground. There's nowhere else I'd rather be.

On this job, though, from the time I got here, there've been rumours of layoffs. Most unions work hard to get what's called "seniority" for their members. That means last one hired is the first one fired, so you always know where you stand. But in the Carpenters' Union, there's no seniority—the boss can lay off whoever he pleases, whenever he pleases. At any afternoon coffee break—which is when it usually happens—the foreman takes you aside, hands you a pink slip and you're gone. The bosses say it keeps people on their toes. So when one of the other carpenters tells me, "You're a good worker, they'll keep you," I know by now that being good isn't good enough if the foreman doesn't like you, no matter who—or what—you are.

That Friday, tension in the shack is higher than usual. The bosses huddle and foremen keep disappearing into the supervisor's shack. That afternoon, five of the crew are laid off. I'm not one of them.

On Monday, a new foreman gets me building fly forms and gives me a new apprentice, Al, who drives me crazy. It's an apprentice's job to do what they're told, but no matter what I ask, Al pretends not to hear, says he can't do that, refuses to take direction. I figure it's my fault, until one day he complains he's never had a good teacher and has hardly ever worked with a journeyman. No wonder, I think, standing up to hand him the end of my chalk line so we can chalk the next few rows for nailing. This is important. The plywood decks will lie on metal joists that have a narrow wood inset to take the nails. To give the form maximum strength and keep it safe, you have to nail precisely. Al drops the line I've given him and walks away.

I stand with my mouth open. Unbelievable! Only when I explain again why we need these lines does he come back. I know most carpenters would never bother explaining in the first place, and sure enough, someone later tells me Al's on the verge of being laid off because nobody else will work with him. When we finish nailing, I ask him to help me lay ply. Instead, Al changes a saw blade as I pack eight sheets of fifty-pound form ply over three plank bridges. As I carry the final sheet past him, he has the nerve to tell me what we should do next.

This time when I blow, it isn't like at Morris, on the framing job. This time I swear at the top of my voice, as I did with Morris, only I'm perfectly calm inside.

"Al," I yell, "you've got three fucking seconds to get your goddamned ass over here and lift the other end of this fucking goddamned sheet of plywood, or I'm going to break it over your goddamned bloody head, so help me Christ!"

The effect is amazing. Al jerks upright, pulls the plywood from my hands and asks eagerly, "Where do you want it?"

After that, I develop an affectionate dislike for the guy. The next time he starts lagging, I swear again, as an experiment, and Al zaps back into co-operative mode. For years, John has told me, "Use your anger!" Now I understand; anger is a tool I can pick up or put down as I wish. And it works.

I've noticed that men tend to pick on me if they sense unease, so I've learned to fake confidence, and more and more I'm not faking it. The banter, the one-liners, the getting tough when you have to, are all just part of the culture. I'm almost bilingual now in this male language, can even appreciate its focus on fewer words and more action. Many of the things I do now—including the way I'm deliberately mean to Al— would have revolted me a few years ago, but now I simply see them as the way to fit in so I can do the work I love. "Fake it 'til you make it," the men say. But you need real skills to back up the talk, a physical confidence on the tools that can't be faked—that's how you win respect, even from the skeptical. All of which explains why Al doesn't get much respect on this job.

After a few weeks of building fly forms with Al and a foreman named Amadeo, who calls me "Kottie," the crane flies our first set of forms up to the top deck, three stories above. Knowing every nail, every notch, every length of plywood, Al and Amadeo and I stop working to watch from below as the crane lifts the form gently into place. It's like a huge, silver-trussed bird. As it lands, three carpenters reach up to guide it onto the deck, their arms in a wide V against the blue sky.

Amadeo throws his hands in the air in salute and, with a rare grin, says, "Kottie, it fits perfect!"

That day at lunch, one of the guys reads us an article about a lawyer who sued an airline for a huge amount of money because his flight was overbooked. All the guys agree, that's a smart guy. But I've seen both sides of this "smart" business, including having two university degrees, living with a law student, and now years of working with men like these who have a high school education yet are brilliant—PhDs—with their hands.

"What do you mean, 'smart,'" I argue. "If it wasn't for carpenters, who'd build their houses for them, their courthouses?"

"Yeah, but they make the money...!"

Finally one man admits, "I told my lawyer once: without the carpenters and the plumbers, you'd be living in a cave and shitting from a stump."

But he and I are in the minority.

—

Tom, the foreman, has begun to act strangely around me. Sometimes he'll discuss with me, as a peer, how to build a certain form, but at other times he's crude. Once, when we're alone, he asks if I want to go watch a dirty movie with him after work.

The other men notice and—surprisingly—speak up. Amadeo tells Tom, "Don't pick on Kottie." Another tears out an article entitled "Men Are the Weaker Sex" to give to Tom, because, he says, "Tom is rude to you." It makes me realize that when I thought Tom's behaviour was obvious only to me, the others noticed too. It's funny; for the first time ever, I feel confident enough that I don't need their support, and here they are, bountiful in giving it.

It's mostly because of Tom that I'm not sorry when the company sends me to another job, building two stations for the new Vancouver Skytrain system. It's the first time a company has ever wanted to keep me, which feels good, though the job gets off to a bad start. Marcel, my partner, introduces himself by taking the hammer out of my hand, saying, "Now, that's not how you should hold that." I don't know who's more surprised, Marcel or me, when I grab my hammer back, put my face very close to his and say politely, enunciating clearly, "Don't you ever touch my hammer again."

He never does, though a few times he reaches for my knife, my pencil, my square. Each time I "accidentally" come down hard with my hammer arm on his arm, until he stops. My tool belt has become an extension of my body; touching my tools without my permission is like touching me.

The one thing they all know about my tool belt is that I always have cough drops that I'll share. It's a solution to a persistent cough, a hit of energy, and sometimes, in the back of my mind, I think it's a peace offering.

The crew on the Skytrain station grows daily, and one of the new carpenters is a Brit named Roger who usually works alone. His steady pace, his confidence and skill, are striking. He must have noticed me, too, because one day when Marcel is away, Roger announces to the foreman, "I'll work with her."

After we lay out the first footing, Roger says, "Good enough," and moves on. Good enough? When I lay my level on the wall, it's perfect. Roger catches my eye and smiles. Roger's and my work together is "good enough" first time, every time. I begin to tease him about it, and about the way he says "Right!" at the end of each job in his drawn-out Yorkshire vowels. When Marcel comes back, Roger says, "You'll stay with me," and I do. The foreman keeps trying to patch me back with Marcel, but as long as you're getting the work done, I learn, or the right person asks, a foreman can be ignored.

I've watched plenty of men at work, but I've never experienced the calm, steady pace or the perfection Roger demands, the way he plans ahead so there are no surprises. I like how he treats everyone on this job the same—with respect. We work hard and fast together. I used to love the talk of framing crews, but now talk seems distracting. It's as if Roger and I can read each other's minds, but it isn't mind reading at all—it's knowing the job and not fighting about who does what. I lose myself in the smack of hammers, the screech of the saw, the deliberate planting of our boots and hefting of materials as we move smoothly, twenty feet above the ground. It feels like dancing. Once I stand back, hands on my hips, to admire our work.

"You going to stand there all morning with an idiot grin," Roger asks, "or are you going to help me get some work done?"

"I'm just admiring the fine work you do, brother."

"You do a pretty good job yourself, sister."

I grin as I slide my hand into the leather pocket of my apron for more nails.

That evening over dinner, a woman friend asks about my day at work, and I tell her it was good. "I banged my elbow on a clamp so badly that for the last hour I couldn't hammer, and my partner, Roger, covered for me."

Her response is shock: I was injured! I should have had time off! First Aid! It makes the job I love seem violent and inhumane. Even my therapist said once, "How can you live with such woman-hating, such misogyny?"

Misogyny? This is one of the best jobs I've ever had. The joy of doing physical work together is one of men's best-kept secrets. If some of these men really do hate women, I can live with it, because I've also seen their vulnerability and their tenderness.

—

Years of unemployment—of not being able to do what you really want to do—kills something in you. I'm terrified about being laid off. We all are. As a crew, we'll do anything to work, pay any price. When the superintendent asks us to give up our afternoon coffee break and instead go home ten minutes early, we say yes. When he asks us to start a half-hour earlier because there's now a night shift, we say yes again. When not one but two business agents come to argue with us, to tell us how hard the union fought to win us that afternoon break and those reasonable hours, someone mutters, "How many paycheques have you missed in the last two years?" Always before, I've felt we should fight to stick to the negotiated contract, but not now. Now I swear along with all the others that I don't want an afternoon break, and mean it. I've long wanted to be a job steward, too, but when several of the guys ask me to run, I let Roger talk me out of it.

"If you're steward, you'll have to stay until this job is over, and they won't be able to transfer you to the next one. Don't limit your chances of work."

Exactly when we need the union most, we're tossing it out the window. Hardly anyone comes to union meetings anymore. After our protest against Pennyfarthing ended with us being told to go home, and Operation Solidarity ended much the same way, the union hasn't felt as relevant. Bill Zander, my hero, has always said, "When the chips are really down, the membership will be there," and "We can't give up. We'll win in the end." But I'm beginning to wonder.

"The union isn't looking out for us any more," the men on the crew say. When I bring it up at Women in Trades, we can all feel it—a slow, scary slide toward the scramble of each one for themselves and may the lowest bid win. As women, we know how hard and lonely an individual struggle can be. I keep going to union meetings, but I don't know how to make things better. I just keep thinking we must.

—

I'm getting so relaxed with this crew I even talk about my personal life, and one day at lunch in the shack, when I mention a potluck, Dave—a wiry little biker of a carpenter who wears Harley-Davidson T-shirts and a Harley belt buckle big as a pie plate—says, "What's a potluck?"

I tell him, and on impulse, add, "Want to try?"

To my surprise, everyone says yes, though some say it more convincingly than others. We decide on a potluck lunch that Thursday. Wednesday afternoon, wondering if I'll make a fool of myself by bringing only a huge salad the next day, I ask Dave, "What are you bringing for the potluck?"

"Potato salad. Just have to add the mayonnaise."

It's rude, but I can't resist. "You make potato salad?"

"Yup."

"Not your wife?"

"Nope. I can make a potato salad."

The next day when the bell rings for lunch, I spread a red-and-white checked tablecloth over the pile of plywood. On it, the carpenters—every single one—lays out food: Dave's potato salad, Walter's fruit salad, bean salad from me, a chocolate cake Al baked himself, and an apple pie from Roger's wife. Our labourer brought a salmon he's caught and smoked. Dave and I provide the plates and cutlery, and in the midst of roaring cement trucks, pumpers, bustling inspectors and three iron-workers who hang off the causeway overhead, watching incredulously, we dig into a feast. At 12:29 precisely, food, cutlery and tablecloth disappear, and at 12:31 we're back at work, looking like your ordinary crew.

—

One night someone breaks into the carpenters' shack. I only lose a tape measure, but others lose more expensive tools, and one guy's boots are gone. Someone is pointing out that the company is responsible for the safety of tools just as the owner walks in.

"So will the company replace what was stolen?" I ask him.

"Why should we?"

Everyone in the shack is listening closely, but no one says a thing. The next morning I bring it up again, and the owner and I end up outside, yelling at each other. "I checked last night and it says in our contract you're responsible for the security of the shack!" He's a very decent guy to work for, but I'm furious.

"And if you aren't happy with how we do things around here, there are two hundred people in the hall who won't mind!"

"That doesn't make it right!"

When I glance around, the crew are standing in a close semi-circle behind me. I appreciate the support, but it would feel a lot better if someone would speak up with me.

"You have to suck up to the boss, Kate," Roger warns me later. "You should have kept your mouth shut." And a woman in my consciousness-raising group says, "You can't do it all. And you can't do it alone."

Still, the next day, all the stolen gear is replaced.

—

We have a new foreman, and he's a jerk. Lou takes his breaks not with the owners and superintendents but in the carpenter's shack, which means we can't complain behind his back. On the first day it rains, he says to me, "I thought I'd see you in a miniskirt raincoat with a big slit up the middle."

"I'll wear my mini when you wear your kilt," I say. Everyone laughs.

Lou tells a lot of jokes, and they're always about sex. One day at lunch he tells one about gang rape. With a sinking feeling, I think, This is it. Lou has just turned me into a sex object. Now these men will start seeing me as a woman, not a carpenter, and make this job that was so good a hell. There's dead silence in the shack. Then Roger pushes back from the table and says, "Why don't we all do something useful?" He leaves, followed by every one of the other men. It isn't even the end of lunch break. For a second it's just Lou and me, and then I leave too, dizzy with surprise. These guys are on my side.

The fact that Lou is a terrible foreman makes it easier for the crew to support me. Lou can't seem to read plans, and he forgets things. We'll start working on Y, and he asks, "Why didn't you do X?"

"Because you didn't tell us to."

"Well, drop that and do it now." Ten minutes later he's at us for not doing Z. And he yells when what we did, as he'd ordered, is wrong.

So I get smarter. When Lou tells me the pattern for snap ties is every 300 millimetres, I make him repeat it in front of two other carpenters. Two hours later when he says I got it wrong, that it should be 320 millimetres, one of the other carpenters backs me up. Still, we all know Lou is telling the owners we're responsible for the mistakes that are suddenly showing up on this job. We find ways to joke about it. One day, when the owner bawls us out for another of Lou's mistakes, he ends by saying, "There are bad journeymen on this job!"

"But good journeywomen?" I ask.

From then on, every time Lou doubles back on an instruction, we all say, "Bad journeyman!" and laugh. Until Dave says it should be, "Bad journeyperson," and we change it. We also give everyone nicknames. Behind Lou's back we call him Louie the Lip, and Great Leader. The labourer who's unbelievably slow is Rigor Mortis. I'm Koff Drop Kate.

One Friday, five guys are laid off.

—

Roger and I remain partners; he's a formidable carpenter. His strength is both in the systematic way he approaches the work and in how he charms his way through a job, starting with the labourers who knock themselves out to bring him what he needs. I follow his lead, never more than a heartbeat behind. Our work together has become slightly competitive, with an edge of respect and enjoyment at how well we produce together.

One day, he casually suggests that when this job is over, we should advertise for finish work together.

"But I'm not a finish carpenter."

"You're a carpenter, aren't you?"

I screw up my courage to ask if he wouldn't rather work with Ron, the next best carpenter on the job, and Roger's friend. "Ron is harder to get along with," Roger says, "and he doesn't do framing or siding." Finishers need a construction carpenter like me to set it up for them. I want so badly to go on working with Roger, it scares me.

All week, rumours fly about layoffs, and on Friday morning, when Lou says, "You ask dumb questions, Kate," I'm sure this is my last day. I say goodbye to my favourite guys, get Roger's phone number. While I wait for the axe, I'm hysterically good-natured, laughing out loud at my newspaper horoscope when it says, "New opportunities arise." But when Lou comes on deck with the layoff notices, he walks past me and lays off two other carpenters. It's like Russian roulette. Four carpenters and a labourer still to go. At the most, there's one month of work left.

It's late October and the days have been getting shorter. It's still dark at 7:30 every morning when we begin work. We've asked Lou several times if we can start later, and finally one morning as we're getting ready in the shack, he announces we'll be starting that day, and from now on, at 8:00. Fine. Except that at the break, he mentions we're expected to work an extra half-hour that afternoon. For a change, it isn't me who asks if we'll be paid for the extra time.

"Why should we pay you?"

This would be easier if we had a job steward, but after the last steward quit, no one—including me—had been willing to take his place.

"We could have been in bed an extra half-hour this morning if you'd told us last night," another guy says.

"There'll be no dissension," Lou snaps. That's his new favourite word. "No dissension or there'll be layoffs!"

Luckily, Lou doesn't come into the shack at lunch that day, and we have a heated discussion. Two guys—a carpenter and a labourer—say we should swallow it, and another carpenter and I think we should down tools and go home at 3:30. Roger doesn't say anything. At 3:25, cleanup time, we watch each other from under the brims of our hard hats. My heart is pounding. At 3:30 I say, "Well, Roger, what do you think?"

Roger looks unhappy. Slowly, because my hands feel stiff and my legs are trembling, I disconnect the saw and start rolling up cords. I'm not going to give Lou an extra half-hour of my time because of his rotten organization. Roger packs up more slowly, and when I look over at the others, they too pick up their tools, and we all head into the shack.

On Friday, I'm laid off. I tell myself it's okay; there isn't much work left anyway and I didn't take Lou's shit. Still, I phone my therapist for an emergency appointment. Only in her office do I dare ask, why didn't I ask the guys to elect me job steward at noon that day? A steward can't be laid off for objecting to unfair demands. Why am I looking after all of them, but not myself?

—

Dad comes to Vancouver and, over supper, tells me stories I didn't know about the family. I knew his father, my grandfather, had been an ironworker foreman, but I hadn't known my grandfather's father was a brass polisher—a lost trade. And Grandpa, it turns out, was a tough dude. Once his entire crew quit because they were in mud to their knees with rain pelting down and Grandpa wouldn't let them go home.

"He'd work in weather so cold, his hands stuck to the steel," Dad tells me. "And he never took a holiday. Your grandmother and I would leave him at home, working, when we went off." The stories make me feel more a part of this family and less like the rotten apple off a twisted branch.

—

Monday morning I check in at Dispatch, but things are worse than ever. Steve's forehead is curled into permanent putty folds. I line up several small renovation jobs for myself, including a couple of days of work for a woman contractor who's new in town, Jacqueline Frewin. She's from Toronto and self-taught. Jacqueline needs help putting a new roof on a big reno, so I call Richard, my friend from third-year apprentice training. We've stayed in touch and occasionally worked on small jobs together. The job goes well and I like Jacqueline's no-nonsense toughness, her care with the work. It's great to know there's another woman contractor in town, especially since Chryse's decided to go back east.

The rest of the Skytrain crew are laid off the week after me, so Roger and I agree that as soon as I finish the small jobs I have, we'll do renovations together with everything split down the middle. I get us our first job. By the end of the week, Roger hasn't been able to make it to the site even once, so I finish the job myself. It soon becomes obvious that his idea of splitting everything down the middle is that I'll find the jobs, do the estimates, arrange the contracting and do the work. Once he's built the cabinets (after I pay for materials), we'll split the profits. When I find out that he's neglected to mention he's also working elsewhere, I tell him our partnership is over.

It's December, and I celebrate the end of my first year as a single woman by having a short, intense affair with Richard. He's leaving in two weeks for a year of travel, so, "Why don't we have a crazy affair before I go?"

We spend most of the next two weeks in bed. I'm dazed by the good sex, good talk and a man who not only brushes my hair and brings me orange juice in bed, but who understands the beauty of rafters.

14

the beauty of ply

At one of the presentations I did on women in trades, I'd met a woman named Joanne Gordon who asked if I'd like to bid on a renovation to her house. After I'd looked over the plans, I talked to her husband, Dave, on the phone. Dave was an engineer, and I talked to him like a guy would talk, throwing in lots of technical terms without explanation, speaking firmly, confidently, with few questions. I figured a guy who knew something about building might also be interested in a new pricing system I'd recently read about called cost-plus-fixed-fee, under which the owner pays the bills as they come in, including labour. For me there'd be a contractor's fixed fee at the end, based on a percentage of the original estimate. It meant less financial risk for me and probable savings for the client. Dave liked it, but first he wanted to meet me, so I'd put on my professional clothes (navy blue turtleneck sweater, clean jeans) and gone over with Roger.

Dave was clearly uncertain about a woman carpenter/contractor, but he was reassured by Roger's presence. Was Roger working for me or did we work together?

"Together," Roger had said.

This will be the biggest job I've ever done, so now that it's time to submit my bid and Roger's out of the picture, I need another carpenter, fast.

I screw up my courage and call Jac. He doesn't want to commit himself in case something comes up over the three months this renovation will take. I call another carpenter I once worked with, but he's busy too. Then a third, who cites a bad back. A fourth man says flatly, "No." Doesn't anyone want to work with me? Then I remember Jacqueline, the woman contractor from a few months back.

Jacqueline says "Yes," just before the first carpenter calls back. He has mixed feelings, he says, but if the job is still open he'll work for me. I know him well enough to ask why so many men have refused to work with me. Might it have anything to do with my being a woman?

"It might," he says. "You shouldn't have said, 'work with me,' you should have said, 'work for me.'" Then, joking, "Maybe it would help if you wore a badge that says 'I'm not really a woman.'"

I'm thinking maybe I should wear a badge that says, I'm really a woman; now be a man and deal with it. But I don't say it and I'm grateful for his honesty. "You sounded unconfident," he continues, "and I was afraid you'd want long emotional discussions instead of just getting on with it. If you weren't confident, a man would have to bail you out, which is okay in some ways," he admits, "because he'd get the perk of being competent. What would be worse is if you were competent, because then a man would feel incompetent or worse—equal."

We say a cordial goodbye, but when I hang up, my hand stays frozen on the receiver. If this is what the sensitive men are thinking, then what about the others? At least it's now clear: the challenge for me on this job will be to take on authority. The challenge will be confidence.

—

The Gordon renovation is an addition, a perfect job involving a little concrete, a little framing, a little roofing and a little finish work—just over ten weeks' work and fifty thousand dollars. Three people are bidding, but one drops out and the other's bid is very close to mine. I've told the Gordons I'll be working with Jacqueline now, not Roger, and they ask the two of us to come over. It's clear Joanne has been on my side all along, but she's letting Dave make the final decision. After a few questions, Dave says, "All right, then. We'd like you to do it."

I can't help but grin.

The job won't start for two months, so in the meantime I work on four smaller jobs, and in the evenings get ready for the Gordons by buying a telephone answering machine and taking several free union courses—on interior finish, door hanging and cabinetmaking. There's a union election coming up so I also go to a long string of meetings about who will be on the Left slate. I'm busy organizing a Women in Trades panel for International Women's Day as well as a benefit for Chryse, who's decided not to go east after all, but to teach carpentry in Nicaragua.

I see Kevin at least once a week. Between going to school, being in the school band, living with his father, seeing his mother every second weekend and staying with me every Wednesday night, the child is almost as busy as I am. He's fifteen now. One evening when I suggest he try a new course at school, he tells me only nerds do that.

"Nerds?" I've never heard the term. So he informs me you're either a nerd or a dude and nobody wants to be a nerd. And he's definitely dude.

He also gives me the lowdown on girls. "Any girl can be your friend, regardless of her looks," he says, "but you have to put on a show in the hall in front of other guys for the beautiful girls. But only for the beautiful ones!" He illustrates with a hilarious performance of the puffed chests and repartee of the boys, then the simpering and lowered eyes and flung hair of the girls, and I watch him, the quiet fire of this child who overnight has stretched into a six-foot handsome teen and a charming, funny companion. Later he says he wants a glossy lipstick because the Duranies like boys with glossy lips.

Duranies?

He rolls his eyes. "Girls who like the rock group Duran Duran!" He says most guys won't tell their parents they want lipstick. And I see that Kevin and I have our own relationship. What he can't ask his parents, he can ask me—his special friend who loves him. Thursday morning, I drop him off two blocks from school so he can walk solo to the front door. It seems only nerds have moms who drive trucks.

—

On the first Monday morning of the Gordon renovation, a backhoe arrives bright and early on the back of a flatbed truck. The morning before, I'd stood, scared shitless, in their beautiful backyard with the surveyor holding an old brush and a can of white paint. We'd staked out the footprint of the new addition, then traced in white paint a line three feet beyond that. Now I had the location for foundation plus drainage space, and the backhoe would just have to dig inside the white lines. Easy.

The backhoe is far bigger than I expected. I've hired a local excavator and drainage contractor named Steve, someone I found in the Yellow Pages, and have spelled out for him a few times that he has to be very careful to avoid the Gordons' prize garden. He'd said on the phone it would take a couple of hours to dig the hole, but after one hour it's clear it'll take longer, so I set up the pump I've rented to keep the hole clear of any possible water and leave to pick up the final cheque from my last job.

"Remember the garden," I remind the backhoe driver again before I go.

With a look of scorn, he turns back to his machine. It's just starting to rain.

When I return an hour later, the backhoe belches diesel over a sea of black mud. The garden has disappeared. There is no grass. Only the backhoe operator and Steve, its owner, standing to one side like a chubby imp.

"Holy shit! I'm dead! What have you done?"

"No problem. I'll get my man to touch it up."

There's been only a light rain, but my excavation is already a small lake.

"Nothing I can do about the water," Steve says. "Your pump's broken."

Calm, I tell myself. Be calm. Think! I phone the folks at the rental company, who immediately send a mechanic to fix the pump. After a few minutes the man says, "You've got a bigger problem here. Your backhoe's broken the main water line to the house."

When I confront the driver—there's no longer any sign of Steve—he turns his back, as if I haven't spoken. Is this my fault? I catch the pump mechanic rolling his eyes, and I know it's not me who has it wrong; this is a mess. I find the main water turnoff for the house, then, since I'm paying him by the hour, send the backhoe driver back to work. Next I call the plumber.

The mechanic can't fix this pump; he'll have to get another. He'll be right back, he says. Meanwhile the backhoe's still digging, and when I check the levels, he's gone too deep. We'll have to make up the difference with more concrete. More money, and the Gordons will be getting a higher crawl space than they figured.

The job Steve's promised would take two hours, takes eight. And when his truck dumps a load of drain rock late that afternoon, the truck gets stuck. In what remains of the yard, the driver spins a muddy rut three feet deep before calling a super-sized tow truck to pull him out. Another fifty-five dollars. The garden is only a memory, and by now this scene of carnage has collected a small crowd of neighbours. Every time I begin to panic, I stop myself; I can't afford to waste time beating myself up. Even if I've made mistakes, I need to be a carpenter and get on with the job.

That night, as I'm about to crawl into the shower, Dave phones. He's remarkably calm about the swimming pool that was once his garden, the ongoing noise of the pump, the broken water line.

"The plumber just left," he says. "Is everything else okay so far?"

I consider hysterical laughter. "It's fine," I say. And cross my fingers. After all, I have a splendid hole—just a little deep—for the new living space, and tomorrow Jacqueline and I start forms.

—

Jacqueline is new to concrete, so I watch her out of the corner of my eye, following up on her work as inconspicuously as possible. Especially, I check her sense of accuracy against mine. When she says, "Right on," I measure later, and it is—right on.

Two guys from the coring company who've come to cut the opening in the existing basement ask what kind of heat we're putting in, and I say, "Electric," then I ask how much it would cost to cut room for duct-work.

"If it's electric heat, why do you need ducts?" Which is when I re-member the heat is forced air and Mickey hasn't said anything about needing openings for ducts. I feel the fog of humiliation move in: how did I dare think I could do a job this big? And catch myself again. I'm al-lowed to make mistakes.

The saw the concrete-cutting guys are using breaks—not once, but twice—and has to be replaced with an electric one that occasion-ally blows all the breakers in the house. When the sand I've ordered for drainage doesn't show up, I phone and am told the truck has broken down. Is someone up there testing me?

Every morning I wake up with the day's to-do list already started on the nightstand beside me, scrawled during the night: "Get extra scaf-folding, 3/8 ply for lintels."

Our forms are built to the book, but every night Dave comes home from work and inspects. Why this? Why that? Jacqueline, who's a lesbi-an separatist with little tolerance for men at the best of times, considers it unbearable. I have more patience. I have to.

"He's paying the bills," I remind her. "And he's curious. He's an en-gineer—he doesn't often get the chance to be this involved."

"He's not letting go of a second of it, either," she mutters.

Jacqueline has attitude. One day, as we lean against the scrap heap in the back yard, eating lunch, she tells me about a man who lived across the street from her last job who laughed at the "lady carpenter" every time he saw her, until the day he came onto her job site and Jacqueline told him, "If I was facing someone with a hammer in her hand, I wouldn't be laughing." He'd come to offer her work, and she accepted.

Jacqueline is built solid, with not an ounce of fat on her. And tall. She has a cap of raven hair and a droll sense of humour that only slightly softens her habit of saying precisely what she thinks. Her voice is as big as her body and pitched low for a woman's. Maybe that's because of the smoking. Once, when I ask how much she smokes, she laughs and says, "Too much!" Her laughter's big, too, as if it comes from deep within her strong carpenter's body.

Part of the plan here is to replace the drain tile around the house, and Steve sends his labourer to dig it by hand. "So it does less damage to the garden," he manages to say with a straight face.

The labourer looks like my grandfather, only shorter, skinnier and more bent. Without a word to us, he bends to begin digging at the southwest corner and doesn't come up until, in an astonishingly short time, he has a hole dug all around the foundation. The hole is precisely eighteen inches wide, three feet deep and as neat as if it had been cut with a knife—a beautiful job.

When it comes time for the concrete pour, I phone Vern. Even with Jacqueline's faith in my abilities, I want to have someone around who's comfortable when liquid rock begins to roll. On pour day, I get up at 5:30 to rent equipment, and by the time Vern shows up right behind the concrete at 1:00, Jacqueline and I are hyped and ready. Using a trick I learned from Jac, I've built these forms so we can pour everything—footings, slab and walls—in a single pour. I've also opted for the luxury of a pumper truck, so there'll be no wheeling of this tricky liquid. All we have to do is direct the concrete into the forms as it pours out of the eight-inch hose.

I work as troubleshooter, checking braces, puddling and scooping extra shovelfuls under his trowel every time Vern yells "Concrete!" Jacqueline handles the hose by holding the tube of gushing concrete between her legs—"easier on my back," she says later—while the man driving the truck studiously avoids looking in her direction. When the same guy tries to tell me how to move the heavy shovels of concrete, I say, "Don't tell me, show me," to get him off my back. I know for a fact that concrete drivers never lift a finger to actually place their product.

But Jacqueline has overheard and announces loudly to the air above her, "Seems to me there are too many roosters and not enough egg-laying chickens around here!"

After that, there's silence, and as Vern begins the slow, careful work of a beautiful finish, we celebrate with orange Slurpees.

—

My estimate for the foundation had been two thousand dollars, but the actual cost, not counting replacement of the garden, is $3,400. When we strip the forms, we find we've poured the concrete a half-inch too high, which means cutting a half-inch off every joist—a huge time-sucker. Mickey, the heating guy, has bid the job sight unseen and now realizes he'll need a hole in the existing concrete wall to join heating systems. It would have cost next to nothing for the concrete-cutting guys to do it; now it will be an extra four hundred dollars and Mickey wants me to cover the cost. Eventually I'll come up with a cheaper solution, but for the rest of that day I can't measure right, and just as we get ready to nail on the two-by-ten header that will hold joists for the new floor, I notice an electrical wire exactly where the header has to go. The cable company says it will cost $37.50 to move it. Oh, and it will take them two weeks to get here.

Jacqueline, only half joking, talks about quitting.

But in the middle of screaming out loud with frustration in my truck on the way home, I think, I'm enjoying this.

Have I lost my mind? No, really, I like this. I like the challenge. And if they fire me, what's the worst that can happen? I'll get more small jobs, I'll build my skills, I'll try again.

That night, Jude and I go to a talk by Andrea Dworkin, an American feminist and the author of *Woman Hating*. It's such a relief to be reminded that I don't have to be nice, don't have to smile, don't have to look after every man in my personal life or in my work. "Be ruthless!" Dworkin says. It's balm, although she also warns, "You must be prepared to watch someone suffer the consequences of their loss of power. They might not be happy about it."

Which comes just in time for meeting the plumbing inspector.

Before rock and dirt can be backfilled over Steve's drainage pipes, Jacqueline and I coat the concrete walls—as per Building Code—with a thick, sticky waterproofing informally called "tar" that will act as a vapour barrier. The plumbing inspector okays Steve's drainage work, then stands with his arms crossed, staring down at the black-painted foundation walls. "You'll have to call the building inspector for your vapour barrier."

"He said you did it."

No. But this is the last day for Steve's backhoe; if the drainage can't be finished today, there'll be an extra charge.

I beg. Couldn't the plumbing inspector check it now and tell the building inspector it's okay?

Well, he might, except it won't pass, he says. "It's bubbling." Hasn't seen bubbles like that in all his twenty years in the trade.

I jump into the ditch to pop one. "Look! Solid tar underneath. It's still waterproof."

In the end, he makes us pop every single bubble and then, after I make a trip to the lumber yard for more tar, repaint all the walls while he watches from the top of the ditch, arms still crossed. A very careful inspection follows, but this time he can't find a single thing wrong.

By now it's the end of the day and Jacqueline's gone home. Steve's man has only half the dirt back in the ditch when I have to leave to pick up more materials.

"Remember," I say to him, "the Building Code says you have to leave an eight-inch space between the final height of dirt and the top of the concrete." This is to keep out insects and water.

"No problem," Steve says.

When I return, the backhoe is putting the final touches to the fill, three inches from the top of the concrete.

"You've overfilled!"

"It's fine," Steve says. Over and over. Even when I bring out my Building Code to show him where it says earth must be eight inches from concrete, he waves me aside: hysterical woman. I know what happened; he had too much dirt, so instead of paying a truck to haul the extra away, he had his man simply heave it on top. To add insult to injury, I'm paying him by the hour.

"No problem," Steve says again, condescendingly. "If you really want it out, I'll have my man dig it to eight inches." Which he does. And sends me the bill.

I talk the situation over with Dave, and we agree on what we should do. When Steve comes by the next day to pick up his cheque, I tell him, "I won't pay you. It was your responsibility. You knew the Code, and I reminded you to leave eight inches."

Steve is a foot taller than me, and heavy-set. "I'm a reasonable man," he says, drawing himself up. He'll compromise; I can pay half now, half next month.

"I won't pay you anything," I repeat. "It's your responsibility."

He moves in closer. He argues.

I explain.

His voice gets louder. He demands. "Full payment—now!"

One of the techniques I learned in a recent assertiveness training course was the Broken Record. "Keep repeating what you want, over and over," the teacher had said, "until they hear you."

So I say it again. "I won't pay you. It's your responsibility."

Steve is yelling now, waving his arms.

"It's your responsibility."

We stand toe-to-toe in the dirt as he roars at me, sweat popping off his forehead. I cling to my script until, finally, he curses me, turns on his heel and stalks away.

I'm trembling when Jacqueline comes out of the house. She and Dave have been listening to the whole drama, my David to Steve's Goliath.

"You were brilliant," she says.

My victory won't be permanent, though. A month later Steve will sue me in Small Claims Court. After finding everything done according to Code, Dave surprises me by taking the morning off work to come with me to court. I know I'm in the right and I look forward to proving it.

I watch as, in a low, confidential voice, Steve tells the judge how he's been thirty years in this business, never a problem, did his best for this woman on her first—"her very first, Your Honour"—construction project. I take the stand and tell the truth. I give the judge my credentials, tell him I've been in the trade for seven years, contracting for two. I have the Building Code in my hand, but when I try to show him the line where it specifies "eight inches," the judge doesn't even glance at it. He leans back.

"It looks to me like you owe this man some money." And before leaving court, I have to fill out a cheque for the full amount. Dave quietly insists on paying me back.

⚊

On the days when Joanne isn't working, she makes us tea at break times. Their three-year-old seems to have adopted Jacqueline and me, and we've both begun to pack treats for her in our lunch bags. One day, the child puts a towel around her shoulders and races in a circle, yelling, "I'm a He-man!"

"You're not a He-Man. You're a She-Woman," I tell her. For the rest of the day she runs around the house proclaiming, "I'm a She-Woman!"

She-Woman's dad continues to come home every night with questions. Lately, he's even started coming home at lunch to see how we're doing. I'm beginning to think he hired us because no male carpenter would have put up with him. One morning on the drive to work, I realize I'm furious at Dave. When I mention it to Jacqueline, she nods.

"He's sabotaging you."

When Dave turns up that day at noon and starts asking questions, I tell him firmly I want to finish this section before lunch. Without waiting for an answer, I go back to work, hammering hard. Dave is nonplussed. He looks around, gets stuttery, then goes over to carry on a long conversation with my dog, so I have to fight to cover a smile. When I call lunch break, I look Dave straight in the eye. Before he can say anything, I remind him that he'd said he'd do chores for us. I pass him the builder's level and tripod.

"Would you take these back to the rental?" He leaves almost immediately. Amazing!

Jacqueline says, "That was much better."

The next day, Joanne apologizes for her husband.

"At work, he's a project supervisor," she explains. "He does it automatically."

Dave apologizes, too. When I ask that he not interrupt if I'm in the middle of something, he agrees, and after that he can hardly do enough for us.

Our hours at work are getting longer. I'm at the lumber yard by 7:00 a.m. and home after 6:00 p.m. to do my share of the cooking and cleaning before sitting down to make a materials list or phone sub-trades for the next day, and always, I'm practising the next day's framing in my head. From time to time, John and I talk on the telephone. We sleep together a few times, but when he tries to tease, "If you don't know how to frame by now, you're in trouble," I take it with not a shred of good humour. I'm trying not to think the same thing. I'm scared stiff about getting this job right.

Jacqueline, too, had heard Andrea Dworkin speak, though in a big auditorium full of women we'd missed each other. In question period I'd asked about femininity: how do you identity as "feminine" when you're breaking out of traditional definitions in every other way?

The next day Jacqueline had told me, "I knew that was you asking— you're such a femme!" She said it in the half-teasing, affectionate way she and her lesbian friends call each other "dyke." Since then, "femme" has become a joke between us, though to me, it's not funny. The question hovers for me like a bothersome insect. As I go through the toughening process necessary to run a big job, I sometimes catch myself looking in the mirror to see if I look any different. One morning I ask the mirror out loud, "Who am I?" like some lost teenager. Yet I feel strong and competent and, in some way, more beautiful. When I catch a glimpse of myself in store windows I surprise myself with how fine I look, but when I come home at night, I'm terribly lonely. I want a lover, but at the same time I know the only thing I'm in love with right now is my trade—besotted, body and soul. Every once in a while I make time for some human connection—dinner with a friend, a Women in Trades meeting, and once I call up a man friend to chat. After he's asked about the job he says, "Men must be terrified of you."

How come just when I start feeling good about myself, the men get scared? The only one who doesn't is John.

—

The new roof will be a simple gable, but since Jacqueline has never built rafters, I buy insurance. I phone Jac, and also my friend Marcia, who loves rafters the way I love concrete. Four carpenters means not only four heads but eight hands to get the roof on fast. I spend the weekend going over my textbooks, and on Monday morning I lay out the ridge and cut and set up the first two common rafters. They kiss perfectly at the peak.

"Nice, eh!" I say to Jacqueline.

"Of course," she says. She never doubted me.

Now that we have a pattern, I go up to the peak while Marcia cuts rafters below and passes them up. Jac catches and hands them up to me while Jacqueline nails the bird's mouths snug to the plate. We pound in one full box of spikes until we run out and I have to buy another, and another, and then we're onto the two-by-three-inch purlins that cross the backs of the rafters, and on which we'll lay the sheathing.

After two years of contracting, I know what I don't know—including the thousands of tiny details in the Building Code. Now Marcia asks me one of them.

"What's the overhang on sheathing?"

I say I don't know without feeling worried I should. Something is changing.

"Jac, what's the overhang on a roof?"

"One inch," he says.

"One inch," I say to Marcia, and go back to working with Jac. He's getting a kick out of calling me "boss." I think he likes the luxury of letting someone else worry, and it's wonderful to have someone with whom I can talk inspectors, costs, clients.

At lunch the next day, before we start to lay plywood sheathing over the rafters and purlins, we sit looking up, admiring what we've done. There's an intimacy to this moment between finishing framing and moving on to enclose the roof. It's the time when a house is most beautiful to me—alive in its bones, pure structure. After the flesh of sheathing and the veins and tendons of electrical wire and plumbing pipes will come the padding of insulation and drywall, and all this beauty will be covered. It's a secret only carpenters know: we are the first to love every house.

Jacqueline is less romantic. When I say, gazing upward, "Isn't it beautiful?" she groans.

After lunch I call everyone over for a consultation on sheathing. The Building Code is clear: a minimum of three-eighths sheathing must lie at right angles to the rafters, so that the maximum of the ply's edges has backing. But we've just laid purlins at right angles over the rafters, and the sheathing will lie on the purlins. So should the sheathing lie at right angles to the rafters or the purlins? After some discussion, all four of us agree on the latter. Jac and Marcia leave, while Jacqueline and I finish up.

Friday, the framing inspector comes. On the jobs I did with Jac, inspectors stuck their heads into one or two rooms, then signed the permit. When one of them made us put extra backing behind a gable wall, Jac complained for weeks.

This inspector is a handsome young man with a somewhat pointed jaw, neatly dressed in a black leather jacket and white shirt. He parks in the back alley, and as I walk forward with my hand out to greet him, he calls in a friendly fashion, "That roof will need more ventilation."

He hasn't even seen the gable-end vents. "The plans were approved. City hall..."

"Doesn't matter," he cuts me off. Standing barely inside the yard, illustrating lavishly on his clipboard, he shows me how the ventilation should have been done. I bite my tongue. If this man doesn't approve the final framing, the job can't proceed.

He then takes out a serious-looking flashlight and crawls under the floor, checking every single floor joist.

"What's with Fishface?" Jacqueline asks, as I stand uncertainly at the crawl space entrance.

"Sssh!"

The inspector's head emerges from the hole in the floor. He takes copious notes during a meticulous inspection of the rest of the house, but says nothing until he looks up at the rafters.

"Why is the plywood running that way?"

"Because of the purlins. I know it's not running across the rafters, but four carpenters, we all agreed..."

"Wrong," he says.

"But we all agreed the purlins change the bearing..."

He cuts me off again with a lecture on plywood grain. "So you see," he concludes, "when you run three-eighths ply the wrong way, your roof is weakened, because there's only a single ply running across the rafters." (Ah. I'd always wondered why it mattered so much, but it's too late now.) He's busy writing something damning on his clipboard. What if he makes me tear the roof off, do it over? I was so sure we were right. There's no way I'm going to let this man see me cry. To justify keeping my eyes down, I pick up a piece of the errant ply, playing with it as he writes.

He glances up, stops writing. "What's that?"

"The ply."

"Roof or sheathing?"

"Roof," I say, puzzled. "The sheathing is three-eighths."

"This is half-inch," he says. "You told me it was three-eighths."

I hadn't said any such thing, but this is no time to quibble. I see now what he's onto. Plywood's strength—which I haven't fully understood until this moment—comes from the fact that each layer of veneer is fastened at right angles to the one above it. Since each ply is one-eighth of an inch thick, you just count layers to figure out the thickness.

I feel a grin threatening to break out. "The architect called for half-inch," I say, feigning confidence. I don't tell him I'd thought it was odd the plans called for more than the standard three-eighths. But the extra ply in half-inch does the trick.

Our inspector has barely signed the final permit, requiring only a small adjustment to ventilation, and walked out the door before Jacqueline breaks into hoots of laughter. I'm shaking even as I laugh with her.

I'd been wrong; we were all wrong about the roof. If it hadn't been for the half-inch ply, I'd have had to redo the entire thing and swallow the cost. I take deep gulps of air.

Jacqueline touches my arm. "You're doing a good job here, Kate," she says. "And Fishface just couldn't believe it."

The electrician who's been working in the next room walks over.

"I've seen a lot of inspections, but I never saw a performance like that. You should report that guy."

"You weren't wrong!" Jac insists that night when I call him. "And if he'd refused to pass the job, you could have appealed it."

When Fishface does his final, finicky framing inspection a few days later, he solemnly shakes my hand and tells me what a pleasure it's been to work with me. I smile. John would call it "not taking it personally."

15

right/left/left/right

Shirley, the architect, has designed me a beautiful business card. "Sisters Construction," it says, in burgundy ink on textured grey stock with my name and phone number. After we finish the Gordon job, Jacqueline and I each go back to our own small businesses, and I'm proud to hand out my new business card as I do a series of small jobs. Having several small jobs is much harder than having one big one. It's the unendingness of it, the juggling, telling your client that what their friend was sure would cost $5,000 will actually cost $8,500 while you grind your teeth, wondering if you could lower the bill for labour—your labour—the only adjustable cost.

I've stayed active on the union's Unemployment Committee, and when one of the union guys asks what it's like to work alone, I say, "Lonely." Apart from the odd conversation with Jacqueline, and an occasional meeting of the Women Business Owners' Association, there's no one else to talk business with. It's one of the reasons why, when my friend Vic suggests I run for trustee in the upcoming Local 452 elections, I consider it seriously. A trustee, Vic says, oversees the union's books, goes to executive meetings and generally acts as the eyes and ears of the members. I'd be at the heart of union business. It sounds interesting. So at the next union meeting, when Vic nominates me, I accept. By the end of the meeting, there are four of us running for three positions. After, a bunch of us go for a beer. Vic is here, and Lorne, the secretary-treasurer of the Provincial Council, with whom I've had long discussions that always verge on argument because we so quickly differ on the best means of getting to the same ends. He's the guy from the union's Left caucus who's tried to convince me that caucus members must always vote together. Most of the guys leave after one beer, but Lorne orders another round and urges me to stay, along with a few others; I'm getting the feeling this is no casual beer.

The bar is the Astoria, a hangout for union people. The Rolling Stones are singing "You Can't Always Get What You Want." I want to dance, but Lorne orders yet another round, and then he's telling me the Left caucus blew it; we should have organized a slate of candidates before tonight's meeting, so now we have the problem of "too many good people" running for trustee. In the next bleary, beer-washed moment, I understand; he's telling me I shouldn't run. As if on cue, one of the other men says bluntly that they want me to withdraw "for the good of the Left caucus." I've never seen the fourth nominee before tonight, but I'm assured he's a solid union supporter who will act the way the caucus decides.

I'm dumbfounded—and furious. When I finally, fully, understand what's going on, I bang my fist down on the table so hard that the ashtray jumps. Every one of the men at the table jumps with it. Who is this "fourth man" I've never seen before? How do they know he'll do a better job than me? I remind them I've been active on union committees, elected to convention...

I'm yelling. One of the guys yells back. Vic, beside me, doesn't say a word. Then Lorne starts talking non-stop, beating away at the idea that I should withdraw my nomination. Weird, the crackle of power, like electricity, though I appreciate that in his own way, Lorne is a straight dealer who'll hang in for a fight.

I learn through what he says that there's another caucus that's Left-of-Left, that meets before the official Left caucus does. I lean across the table and say with beer-induced confidence, "I come from the women's movement, which has a very different style of organizing. I have skills that could be valuable to this union, but you're telling me people don't want them."

"I appreciate that feminist style," Lorne's son Carey says, "but it's one I'm completely uncomfortable with."

An honest man. But all this is crazy-making and I am drunk. I get up to leave and am grateful that Vic leaves with me, because the moment I step outside, I burst into tears.

"You're not alone," Vic says. "We'll find a place for you. We'll fix this up."

"Who's we?"

Which is how I find out that Vic belongs to still another, different, Left-of-Left caucus from Lorne's. How many of these caucuses are there? A few days later, when the Left caucus that I'm a member of—I think—passes out its slate of delegates recommended for election, my name's not on it. I get stubborn. I want this nomination.

I start to talk to everyone I can: business agents, the president of the local, the dispatcher, the head of apprenticeship, other union members—even the mysterious fourth candidate, who tells me he doesn't really want the job. I talk to people not just to get information but to let them know I'm still here. This, too, I realize, is a feminist way of politics. What helps most is talking to the women in Women in Trades and the union women I've met through Solidarity. After a few conversations with them, my question switches from, "Why aren't I acceptable as trustee?" to "Why is this process so controlled?"

When I'm not even invited to the next Left caucus meeting to discuss the election, I feel defeated. I talk to my therapist about dropping out of union politics entirely, but after an hour I find my anger, and that makes me stubborn again. I've always gone to the Left caucus meetings, so I'm going to this one, too—invited or not. And I'm going to run in this election with or without their support.

What upsets me is that the man they're so keen to support as trustee instead of me has always been identified as supporting the Right slate. But because, they now tell me, "he stood out at the last few meetings," they're welcoming him like a lost sheep. For four years I've been standing up at union meetings. I'm an active member of the Apprenticeship, Unemployment, Women and Human Rights committees, as well as representing the Local on the BC Federation of Labour Women's Committee. No one mentions the three conventions I've been elected to, or picketing at Pennyfarthing, or my Solidarity work. I like the new guy, but I hate the contradiction: on one hand I stand out like a red beacon, and on the other, when it comes to important issues like power, I'm invisible.

I take two weeks off work to campaign. It's against union bylaws to campaign inside the union building, so I hang out with the other candidates in the parking lot outside. The split between Right and Left in the local is intense. The Right's slogan is "Time for a change," and their man running for president, Mitch, is someone whose obvious care for the union has impressed me in the past. Once he'd been my foreman on a union job, and a good and a fair one. One day in the parking lot Mitch says to me, "The reason we didn't pick you for the Right slate is that we picked the best men for the job."

It's such a kick in the gut—so unexpected, so personal—that I actually bend over and clutch my stomach. Still, I manage to say, "Funny, I thought your motto was 'Time for a change.'" The men around us laugh.

Before the Left caucus meeting, I am very, very nervous. I pin new affirmations on the fridge: "I will be clear, calm, concise." Vic suggests I talk to John, a guy from Vic's Left-of-Left caucus, which he tells me is called the Canadian Party of Labour (CPL). John (of the CPL) gives me some useful pointers. "Stress that you're not trying to create disunity," he says. "Be honest. Remind them of how hard you've worked on union committees."

At the meeting, knees shaking, I give a speech that's as clear as I can make it. I talk about my commitment to the progressive side of the union, even if my name hasn't been selected for the slate, about how four people running is more democratic than a win by acclamation. I promise the brothers that I'll run because I was nominated, but I won't campaign or in any other way act to divide the local.

In for a penny, in for a pound, my mother always said, so I also speak about the racism and sexism in our local, how it divides and weakens us. I don't think it's a coincidence that I, the sole woman, have been nominated by Vic, one of the few men of colour in the local. I tell them about Mitch saying the Right had picked "the best men" for their slate. The guys find it much funnier than I did. But I'm surprised and relieved when a few of the leaders in the room speak up in support, including Angelo, one of the significant number of immigrant carpenters in the local. Even Lorne half-smiles at me after as we all go out to the bar for a beer. One beer. No hidden agenda.

—

I'm not elected as trustee, which is fine. I feel good that I tried, and shortly after, I'm asked to help organize a small company of unemployed union carpenters to do renovations and small jobs. We call it Rank and File Construction. Some of us are excited about how one day it might grow, how we might actually have a union-run construction company—perhaps a co-op—because around this time I'm also approached by a group of high-powered women who call themselves WomenSkills. They want to talk about us working together, union and co-op, to form a housing construction company. And why not? We have all the skills that the very best companies want.

In my own company I have enough work that I can risk losing some jobs, maybe even save a little money, so I raise my labour price from ten to twelve dollars an hour. When I hear about something called the Federal Business Development Bank that offers courses to small business owners, I take several, including accounting and time management. Whatever I'm doing, it seems to be working. I get 95 percent of the jobs I bid on, though I'm rarely the lowest bidder. Lately, being a woman has almost become an advantage. A lot of people like the idea of supporting women in the trades. One woman tells me she feels safer giving a woman the key to her house, and when I explain to her in some detail why I'm putting an extra bearing beam in her floor after her neighbour was sure she didn't need it, she says that's another reason she likes women—we take the time to answer questions. Kevin, my stepson, asks if he can work for me and over the summer break becomes the company labourer.

In spite of my higher labour fee, money is always tight. I'm on full overdraft at the bank, my Visa is at the max, and still I owe my roommate fifty-five dollars for last month's groceries. My accountant tells me that after basic living expenses of food and rent, I made four hundred dollars last year.

"You're not charging enough," she scolds. "And getting 95 percent of your bids proves it. You need to charge 30 percent overhead." But as soon as I think of "profit," I think "profiteering," and how everyone in the union will hate me, though on the jobs where I know the owners are better off, I now hold my breath and charge what I think my labour is really worth. Any spare money I continue to invest in tools. The natural next step would be for me to expand my business, hire a full-time carpenter and labourer as crew, charge more. But my values are all in conflict. The old Sunday night battles with my father have come back: how I hated what I saw as his scorn—management's scorn—for ordinary working people. So here I am: an ordinary working person and one of those management types, like him, that I'd sneered at. It's complicated. This same dad once told me that whenever he inspected one of his company's textile mills, he'd always park in the employee lot so he could walk across the factory floor and talk to the operators.

"Why?" I imagined some diabolical spying intent.

"What's the point of asking management why the new machine isn't working?" he'd said. "You have to ask the man who runs it. He knows it best."

In the end, I decide I'd hate it if I was solely a manager. If I can't share my business, I want to be a worker with other workers. Which is maybe why I stay busy in the union with meetings three Mondays and one Wednesday of every month.

One night I have a long talk about the union with Liora, my ex-thesis adviser who's now a friend. She's one of the few people whose objectivity I trust when it comes to union politics. "If you were a man, they'd be grooming you for leadership," she says, but I'm not, and they're not. As we talk, the picture becomes clearer. I ran for office on impulse, without considering the consequences. I was attracted to the idea of becoming part of the power structure, but I'm not the kind who fits easily, it seems, into existing structures. Liora says I have three options.

First, I could start a long-term strategic plan to build my own base of support. This would mean at least two more years of low-key work in the union, working on committees and eventually finding others who would run for office with me, on our own slate.

My second option is to be a maverick, "a gadfly," she says, nipping at whoever's in power, calling issues as I see them, stating what no one else dares state. This, too, can be a source of considerable power. Or third, I can just go to union meetings and socials, have a good time and take work when there's work available.

She and I don't discuss where my own business fits, or whether I can be both a manager and a worker, or the union's idea of a business. But when I think later about whether I want to work at forming a power base in the union, I decide the answer is no.

—

That summer I break my own rule and get involved with another construction worker—Mickey, who does my sheet metal and heating work. When I tell Jacqueline, she rolls her eyes: not because it's Mickey, whom she likes, but because she's always thought I should show better taste by sleeping with women.

Once, when I was being interviewed on TV about women in trades, the interviewer had leaned forward and asked, "Is it true there are lesbians in construction?"

"Yes," I'd said. "There are lesbians everywhere, even in television," and she'd changed the subject. But of course she was right; there are lesbians in construction, and the fact that some of us are and some of us aren't has often created division, so at a conference organized by the newly formed Women in Trades and Technology National Network, I attend a workshop aimed at helping us—lesbians and straight women—understand each other.

The exercise the trainer gives us—asking us to place ourselves on a line where one end means we sleep only with men, have only ever wanted to sleep with men, and the other means the same, but for sleeping only with women—amazes all of us. When I look around to see who's standing where on the line, women I know as lesbian are standing closer to the male-only end of the line than I am. Women I know as entirely straight are nearer the woman-only end. The trainer explains what by now is perfectly clear: you can never know from the outside how someone feels about her own sexuality. And it can change.

"Next week, next year," the trainer says at the end of the discussion, "you might put yourself in a different place on this line altogether."

Later, as I discuss this with another straight woman, she tells me about the Tingle Test.

"If I'm sitting on a park bench," she says, "and a beautiful woman walks by, I look. If a beautiful man walks by, I tingle. That's how I know."

It all confirms my love of women: our matter-of-fact courage, our ability to be vulnerable, our toughness and our compassion. I love women's good sense, our pride and our humility.

These days, Jacqueline is my closest companion. We contract our own jobs, but hire each other when we need a hand. Negotiating with Jacqueline is easy. I want to start work at 8:00 but she's no good before 9:00, so we start at 8:30. We'll discuss how to tackle a difficult corner and quickly agree on the best way, then build it. Construction is a wonderful vehicle for getting to know someone at a relaxed pace. Once, as we take a break from digging someone's drainage ditch—we know now how to do it ourselves—I whine about something a man said to me on my last job. Jacqueline stops work to look at me, plumps the hand not holding the shovel onto her hip, and mock-frowns.

"Ka-ate!" She pronounces my name in a slow drawl with a sigh that's half exclamation, half exasperation. There's only a hint of humour behind its steel. "When will you figure out that this is your territory as much as theirs?" And goes back to work.

What a radical idea!

—

The news is stunning. The International, the head office of the Carpenters' Union in Washington, has cancelled the results of our Vancouver local's last election. "Suspected irregularities," they say, though nobody seems to have the details on exactly what that means. John from the CPL caucus phones me. He's in a different local of the union, so he can't attend our meetings, but he comes right to the point. "What are you doing about this election?" He pushes me to run again for trustee. He even suggests the outlines of my campaign. "Focus this time on organizing. The issue now is autonomy from the International. We'll be aggressive."

"I'll only run if Vic nominates me," I say. I won't be a sacrificial lamb, won't stick my neck out the way I did before. Vic rarely speaks at meetings, so there's a good chance he won't speak up at a nominating meeting, either. If that's the case, I'll vote for the slate that the mysterious Left-of-Left have selected ahead of time, come home and carry on with my own business.

The room in which we meet is packed with about 150 men and me. When the proposed three nominees for trustee are read out, I'm not one of them. The secretary-treasurer, who's chairing the meeting, explains the first man has been nominated because he was on the original slate and won, the second (the previously right-wing fourth candidate), because he's been "a stand-out" at the last three meetings. The third is Carey Robson, Lorne's son. But damned if Vic doesn't stand up and give a very nice speech about how he thinks I should be the third trustee.

And slowly, several of the ordinary guys—the guys Bill Zander calls "the backbone of the union," who come to every meeting, who are moved not just by good wages but by the principles of fairness the union represents at its best, guys who've never said two words to me though they've seen me at meetings and on committees, maybe even voted for me to represent them at conventions—one after another, several of these men get to their feet to say things like, "Some of the guys out there don't speak English so good, and they might see Carey Robson and figure it's Lorne's middle name."

I'm almost in tears. They want me to run! Seeing no further speakers, I stand and talk about my campaign strategy, to show them that I have good ideas (well, that I recognize someone else's good ideas), and that I'm not some weird witchy woman.

"I have no trouble not being on the slate," I finish, "but I do object to the fact that the executive—even though they're elected by us—have presented us with a done deal." It's the only time all evening that everyone claps. The chair then suggests we go through the slate one position at a time to decide who'll represent us, and I'm officially elected to the Left slate. I'm delighted.

Suddenly, every leader in the union wants to take me out for lunch. They ask a lot of questions, give me insider information. They want to know seemingly insignificant things like, who do I talk to? As if I have a secret and they're trying to find it out. John from the CPL caucus says they just want to know where I'm coming from. Several of them stress to me—again—how important it is to abide by caucus decisions. But I don't want to belong to a movement, a party. And I hate secrets. I just want to say what I have to say. They seem suspicious of a carpenter running for executive who also runs her own business, and it's unusual, I'll admit, but these are unusual times, starting with 55 percent unemployment. I wouldn't admit it to anyone in the union, but in a way, they're right to worry about me: because I'm getting the employer perspective, I'm opening up to options that aren't traditional union tactics. They see everything as either management or labour—black or white—and I don't feel comfortable with the simplicity of that. At best, they're honourable, principled men who seek respect for everyone, including working people. At worst, they're stuck, old-fashioned, leading us down in flames with glorious traditions intact. They're right to worry—I question the traditions.

Again, I take two weeks off work to campaign. Some of the business agents invite me to go with them to job sites "to talk to the guys." As before, I also spend a lot of time with the other candidates in the parking lot outside the union hall, going from circle to circle of members. We call our slate the Action Slate, as opposed to the Prince Slate, now headed by Jim Prince, so we can avoid the old Left/Right categories.

I tell everyone I'm running because we all need to get more involved, to speak up. At first, I find it scary to be approaching men out of the blue, saying, "Can I ask for your vote?" But I'm not alone. There are usually several of us out there from the Action slate.

When we started, I was physically afraid of Jim Prince. He's tall, rough and heavy-set, and I sense a meanness to him. But one afternoon, when I'm feeling cheeky, I approach a small group talking with him and Mitch, who's now running for business agent. When I confront Jim nose to nose, I discover I'm more articulate than he is. He keeps backing down, losing his place, and finally he walks away, saying, "You can't argue with a woman!"

"Especially when she's right!" I yell at his back.

The Prince slate are campaigning on the slogan, "Choose the right men for the job." Since I'm the only woman in sight, it's hard not to take it personally. After Prince walks away, Mitch says to me, in front of the remaining carpenters, "And who do you represent, Kate—all two women?"

"I represent carpenters," I say.

"You've worked for me," he says. "You're no carpenter." He follows it up with, "Be honest with me, Kate. Wasn't I the most pro-union foreman you ever had?"

I tell him yes, and say I've told people that, but I ask for the same honesty from him.

"Okay. You're a carpenter," he admits. "That was a cheap shot."

One day, Jim snarls at me, "What did you do when your old man dumped you?"

Suddenly I can't breathe. But he's not finished.

"Didn't anyone tell you they always dump you after you put them through law school?" As he walks away, the carpenter beside me sees the look on my face and takes the leaflets out of my hand.

"Go wash your face," he says, and I go into the ladies' bathroom to cry.

I hear the gossip from my side, too. The Left-of-Left guys call me "a loose cannon" and "out of control." But I don't dare ignore my intuition. I have to go by that, even if it means I don't fit anywhere.

A few days later, when there are several guys from our slate on the parking lot, I walk into a circle where Jim Prince, Mitch and several members are talking and start to hand out pamphlets. One man hands his directly to Prince, but I keep on talking, telling them why they should vote for us. While the others continually interrupt, one man in the circle listens, keeps talking to me as the others drop away. He asks good questions, the ones I know everyone is asking: Why isn't there any work? Why don't members come to meetings any more? What's the union doing about it?

"I feel personally shafted," he says.

He and I talk for twenty minutes. When the interruptions of the Prince people become too intense, when my knees are shaking and my heart pounding too hard to stand it any longer, I leave to go sit down in the hall.

The man follows me.

"Thanks," he says, as he heads into Dispatch.

⟶

Our slate wins every seat we run for, and I win as both a District Council delegate for my local and as trustee—the first woman elected to Local 452's executive, I'm also voted in as a representative to the BC Federation of Labour Women's Committee, and to the next Carpenters' Convention. It's almost scary, the sense of obligation it all bestows, but the men don't exactly let it go to my head. After a union meeting, over beer, one of the guys says, "If you ever ran for full-time office in the union, lots of guys would vote for you, for the novelty." And another, "If you run, some of the guys will think you're too damn smart for a woman."

At Women in Trades, we laugh about how I'll act at my first meeting of the executive.

"What will you wear?" someone teases.

"Don't cross your legs!"

"Sit up straight!"

They're happy for me, for the victory this represents for all of us.

At the first executive meeting I do, indeed, sit up straight. As fifteen or so of us gather around a long table in Local 452's union hall, we're a bit formal. But there's an exciting energy. One of the new trustees asks for information on how a sum of money was spent, and I argue for keeping the Unemployed Committee. When there are differences in priorities and tactics, we can sense the old guard yield, a bit grudgingly, but I think we all feel the sparkle of something new.

There has been one unexpected change: since the election, men and women both inside and outside the union react to me differently. They defer, as if overnight I've transformed. For the first time I feel the distinction between personal power and the power of a position. If I decide to go further in union politics, I'll have to be careful: the day I start valuing the position for the perks and prestige it gives will be the day I become what they call in the unions a "porkchopper." Vic tells me I don't realize how much influence I have in the union, how much people respect me. I fear these compliments, maybe afraid that if I take it all seriously, I'll become arrogant—the thing I most hate in my father. Still, for the first time in my adult life, I can feel my own authority. When several men on the Left call me "bourgeois" because I run a renovations company, I feel guilty until I find out two other guys in our caucus have companies big enough to be incorporated. I stop worrying about being bourgeois and start worrying more about my union.

Most of our members are over fifty and some of them are planning early retirement because of the lack of work. A lot are out contracting, like me. But where are the new ones, the apprentices? In October, some of us work with a lawyer to formalize Rank and File Construction. But although unemployment in the building trades is over 55 percent, the executive carries on as if there's no crisis, spending huge amounts of time discussing things like whether we should buy a computer for Dispatch. When I suggest we hold a conference for members on unemployment, the guys around the table stir uneasily.

"What would we say?" someone asks.

"Shouldn't we wait," someone else says, "until we have a plan of action to lay out for them?"

"But that's the point," I say. "Every idea we've tried, including the union company, has hardly made a dint. It won't be us telling them, it'll be us listening. We'll brainstorm. Maybe they have some good ideas of their own."

It's a feminist approach, I know, but I stop talking because of the shock around the table.

"We can't tell the members we have no ideas!" someone exclaims. And we move on.

The more executive meetings I go to, the more clearly I see the split between Left and Right. We don't fight the employer any more, or the government—we fight each other. Still, I drag the secretary-treasurer of the Provincial Council to his first employment equity conference; I ask the executive for more support for the Unemployed Committee and for Rank and File Construction; I press the president of our local to invite one of our members, Phil Vernon, to sing his songs about being a carpenter at our next convention.

In November, we get into a particularly divisive issue at an executive meeting. A right-wing local has made what I think is an excellent suggestion for a way to deal with our lack of work, but the decision comes down to the fact that it must be a bad idea because it was suggested by the Right. The final vote is me and the right-wing guy for, everyone else against.

When I get home, pissed off, there's a telephone message from a friend whose public service union is advertising for a communications director. He wants me to apply. In a way, it's my dream job. It would pay me to write for an organization whose principles I believe in, I haven't had a steady job for years, and the salary he names is huge—forty thousand dollars plus a car. But what about the guys who've just elected me? What about the work to be done for carpenters?

My Women in Trades friends urge me to take the job.

"We're all in survival mode," they say. "We have to keep an eye out for foxholes that will see us through."

But my passion right now is for the hammer, not the pen, and my trade wins.

Nonetheless, I'm beginning to see how running a business leaves me soul-dead. There are perks—being on my own schedule, free to work twenty hours a day, and I'm good at the work—but being good at something doesn't make it right for you. And I miss my friends. I've been focussed 100 percent on work, then 100 percent on work and the union, for too long. I've lost friends, taken them for granted, acted as if the time I spent with them was another chore to cross off my list. I'm not proud of this. I've gone to a few meetings of a group called the Adult Children Of Alcoholics, where I learn I've been living in the past and the future. Now I want to slow down and live in the present. But it's shocking, like a train that's been shooting down the track trying to slam on the brakes. At parties, friends have me crawling under tables to tell them if things are well built. I used to be angry that no one took me seriously as a carpenter. Now I'm angry that they seem to do nothing but.

When Jacqueline phones to say she has a two-week job coming up, doing a renovation for two women on Salt Spring Island, I'm relieved. From now on, I'll do contracting only when I absolutely have to.

Late in December, when I fly home for Christmas with my parents and youngest sister, a neighbour makes a casual comment that my sister looks like my father, but I don't. It's as if a padlock has been opened. All my life everyone has said I look like my father, act like my father—but with that comment, I feel like I've suddenly gained my freedom. I'm not my father. For the rest of the holiday I'm more peaceful with my parents than ever before.

There's another new development: John wants us to get back together. "Once in a lifetime," he says about our relationship, but I'm terrified of going through that pain again and refuse to see him more than occasionally. Still, he's my best friend. When I told him about the union communications job, he'd asked, "What's your ambition?" As usual, putting his finger on the core of it: what is my ambition?

The Salt Spring Island job with Jacqueline goes well. It's fun to just be a carpenter again, working for two women who cook us marvelous meals and are wildly appreciative of everything we do. I'm only a little worried that I have no work lined up when I get back to Vancouver, and one day when I get out of the bath and run the towel over my body, I notice the rounds and the sharps of me aren't so sharp anymore. Where once my ribs stood out clean as sculpture, they're now soft, as if steam has fluffed my edges. Time to go back to work.

I've always kept in touch with Ray Hill, the man who hired me on that very first construction job on Pender Island. One day when I visit him and Beth on Salt Spring Island, Ray asks if I'm busy these days.

"I need a carpenter," he says. "Up north."

16

dease lake

On a Monday morning in mid-April, a friend drives me to the ferry terminal, where I'm to meet Ray's truck. Ray's son, John, has just moved to Dease Lake, a town in northern BC, to start a flying business, and we're going to build a house for him there. The details are a little fuzzy. I've brought my tools, and Ray's paying $10 an hour plus room and board. I was getting $25 an hour plus benefits in the union, but this isn't about the money. I'll build my first house from scratch, make some cash, and get a chance to think about where my life is going—that's the main thing.

"We'll be out of there in four weeks no matter what," Ray had said in his best hearty voice.

Dease Lake is very near the Yukon border, seven hundred miles north of Vancouver. Ray had casually mentioned there'd be snow on the ground when we got there.

There's no mistaking him when he comes. Not one but two over-loaded flatbeds waddle off the ferry, billowing under gleaming orange tarps that make them look like over-heaped bowls of Jell-O. They roll to a stop in front of me, and Ray and a younger man jump down from the smaller truck. Ray introduces me to Dave, a thin man with dark glasses, who will be our labourer. So we'll be a crew.

Ray's driving. I climb into the truck beside Dave, insisting on the window seat—none of that cozy boy-girl stuff. The other truck, the bigger one, is being driven by a friend of Ray's. Off we roll, but after three flat tires on the big truck before we've even cleared Vancouver, Ray stops our parade, phones around and finds a driver by the name of

Beecroft who, for $1,400, will drive our load to Dease Lake in his much more professional-looking rig. Dave dubs the man "Captain Beefheart," and Ray confesses he's worried the guy will take the seven-hundred-dollar cash advance along with our building materials and bugger off to California. After we've made the arrangements with Captain Beefheart, I call my roommate.

"Where are you?"

"I'm just down the road," I tell her. "Bringing guests for the night."

The next morning we transfer all the materials and finally leave Vancouver late in the afternoon. I ask Ray to please stop for the night in Williams Lake so we can all get a decent rest, but for all the next day and the second night, we drive north. By the afternoon of the second day, everything has taken on the same damp grey-green colour, the same lulling rhythm. We talk a little, listen to music a lot. Ray has brought several tapes, all jazz. Whenever we pass within range of a local station, Dave frantically pushes buttons on the dashboard. "Country and western?" he asks hopefully, while I pray for Peter, Paul and Mary. Every four hours we change drivers, and every few hours we stop at a café that looks just like the last one. Each time we stop, the air is cooler, the sky more vividly blue and, as night falls, more intensely spackled with stars. On the second night, I get the 1:00 a.m. to 5:00 a.m. shift. At the village of Kitwanga, I pull the wheel right—north—where the sign says "North to Alaska," and as the two men sleep beside me, I follow my headlights, carving out a tunnel in the dark. With a crescent moon rising, I sing softly to keep myself awake, to keep from being scared.

There's only one skinny road—I've seen the map—but by 3:00 a.m. I'm convinced that somehow I got turned around; I'm going the wrong way. I almost wake Ray. But I hold on, keep driving, keep singing, until to my right comes a slow dawn—"East!" I assure myself—and out of the darkness, like a slow fade, emerge gravel and rocks and trees. The spruce and hemlock here are skinny as toothpicks. Ray takes the wheel and I sleep.

When I wake a few hours later, we're at the bottom of a narrow valley surrounded by low hills, and except for the thin rope of road, we're in the middle of deepest winter. The air smells like fizzy pop. Ahead of us, a metal bridge shines black against a white hill. The road we're on—the only road there is—crosses the narrow arm of a frozen river, then twists back on itself and zigzags up: black on white.

A few minutes later, Ray says, "This is Main Street," and I laugh at the joke because except for a log apparition that says "AM CAFE" over its door, we're in the middle of what looks to me like wildest bush. Four weeks later, after my eyes have learned to distinguish buildings—a pub and a clinic, two motels, two food stores, the post office, two gas stations (one called North Dease and the other, a hundred feet south, South Dease), a bakery, a school and a community hall—Dease Lake will look busy. Right now my city eyes see nothing except trees, snow, and three blue Ford pickups parked in front of a café. Inside, the restaurant smells of fresh coffee and toast, its log walls lined with naugahyde booths over which red-checked lamps hang low. I slide onto the padded seat across from Ray and Dave.

Halfway through the bacon and egg special, a big, rough-looking man walks in with a huge smile slapped across his half-shaved face. Dark hair spews like a fountain above a blue plaid jacket and grease-stained blue jeans. He heads straight for our table.

"Hi, all!" he booms.

"Meet John," Ray says, "my son."

"How was the trip?" John bellows. "I expected you yesterday! May I?" He slides in beside me, wipes a huge hand across his face and yawns— a grandiose affair that involves his entire head.

Teddy bear, I think to myself.

"I ordered a backhoe for the excavation," John says, digging into a breakfast special with white toast. He looks at his watch. "He'll be there right now."

With a snap, my head reconnects to my body. We'd made a deal on the road, Ray and I, that if we drove all night, we'd take this first day off to rest and unpack.

"I thought you were going to have the excavation ready," Ray says, smearing a special order of peanut butter on his toast.

"The guy was busy," John replies. "But guess what! I got you luxury accommodation. Sandy and Al at the motel will give you a deal on a room or two."

"We'll take a look," Ray says, his face deadpan.

The motel, half a block away, is two rows of aluminum trailers facing each other across a courtyard of trampled snow. Each trailer is raised on posts above the ground, and a series of short stairways lead to a common balcony that runs in front. A huge silver fish flips on its side above the unit closest to the street. "Grayling Motel" is written on its side, in pink and green.

The owner, Al, shows us a unit that's a small sitting room furnished with a beige couch and a small Formica-topped table. Around the table are three chairs clustered beneath a small window. Along the right-hand wall stand a stove, sink, kitchen cupboards and a small refrigerator. A token wall divides the kitchen/sitting room space from the bedroom—two single beds covered in red plaid. A narrow door leads to the bathroom—pink walls, pink sink, pink toilet and a pink shower just big enough to dry yourself in without hitting your elbows on the walls.

"The couch pulls into a double bed," Al says.

I've been sort of hoping for a separate room, but when I glance at Ray's face I see that's not going to happen. I drop my pack and turn to fetch the other bag with my boots in it.

"I'm going to have a nap," I announce. Damned if I'm going to look at a single bloody clump of dirt before I close my eyes for at least one hour.

"Good idea," Ray says graciously, but I don't look at him. I know he's ready to go straight to the site and begin. "I'll go over with John and check out the backhoe." There's a masculine shuffle toward the door, then Ray turns back. "You stay here, Kate. Straighten out the groceries and make a little nest for us." And for the first time all morning, he smiles.

Maybe I lost the "nice girl" bones in my body somewhere back on the highway. Maybe they're only resting. But I proceed to make it perfectly clear that my job here is not as wife or nest-maker, and while we're on the subject, they should understand I'll be doing my share but I will not be cooking every meal. No discussion.

After a moment of silence, Dave decides he'll stay behind and help me fix up the nest. Ray and John make a quick exit. As Dave and I unpack, I make a few forays into how he thinks we should set our working limits here. Does he care how many hours, how many days a week we work? After all, he and I are the two outsiders on this job. Then he tells me he's Ray's son-in-law. So much for solidarity.

We nap, me on the pullout couch and Dave on one of the single beds, snoring.

Three hours later I wake, startled. The room is filled with a hollow silver light, and I slowly reorient. I'd been dreaming—what? I was excavating around an ancient house and had just dug up the statue of a hawk, so dark green in colour it was almost black. Its name was Horus. The statue held an ancient power. In my dream, I rubbed against it like a lover while someone told me that because I'd found it, I could take it home. Its power, the voice said, was now mine.

The trailer is very cold. I fold up the bed, then dress in that strange crystal light, in my down jacket and arctic boots.

I feel surprisingly rested when I tiptoe out the door with Dave still snoring in the next room. Outside, low grey cloud threatens more snow. At the road, I stick out my thumb and a woman in a green Ford slams on the brakes, spitting gravel as she stops beside me.

"I'm going to the airfield," she says, and I climb in. She has sandy blond, curly hair and a pleasant face lit by beautiful large brown eyes. She wears no makeup. I explain I've come to build a house for John Hill, at the airstrip.

"A woman carpenter!" she says. "Wait 'til I tell Ron!" Her name is Jan, and she lives with her husband, Ron, and their three-year-old daughter in Telegraph Creek, seventy miles from here. Ron is a pilot, like John.

"That makes him an important person," she says, stating a fact.

At the airport, Ray and John stand on opposite sides of a large hole watching a backhoe at work. It's a bigger excavation than I expected— bigger than they expected, it turns out, for the soil is mostly sand that keeps falling in as the hole grows. Ray has shown me his design—a simple but attractive structure, twenty-four by forty with a straight gable roof. We'll build it in two halves so the house can be picked up—like adjacent trailers—and trucked to another location if necessary. Ingenious.

Jan pitches in to help us pile supplies in John's shed. When the truck is empty, Ray suggests we all go out for supper.

"Where to?" I ask. Jan looks at me strangely.

"There's a good place in town called the AM," Ray replies with a straight face, and back we go to where we started. The AM is the only show in town. If couples want a date or a private dinner, Jan tells me, they just sit in a booth with their heads together "and don't look up." She tells me the story of how she first came to the north, dropped into a snowed-in ranch by helicopter with a man she'd never met before to spend the winter looking after horses.

"It got a little touchy toward spring, when we ran out of food and had to live on rice and beans for two weeks, but other than that, I loved it. Isolation gives you time to think."

I look at her with the same awe I once felt when, on a trip to Spain, I woke up to find oranges growing on trees.

While the men talk airplanes and trucks, Jan tells me she's taking a little break from Telegraph Creek, staying at the same motel as us a few doors down. Once we get back there, I undress in the bathroom.

I've brought my thickest nightgown to serve as nightgown, bathrobe and blanket. When I slip out of the tiny pink bathroom for my first night in Dease Lake, the men are already asleep.

—

In the morning, we each cook our own breakfast and pack our own lunch. As we pull on down jackets and arctic boots, Ray announces, "I'll take first shift cooking tonight."

"Me next," Dave says.

"Me third." I'm grateful they took my outburst seriously.

"John will sleep in the cabin, but he'll probably eat suppers with us," Ray says. "He can take fourth shift."

When we arrive on site, it's below zero. I set up the builder's level, but as the morning wears on, I can't for the life of me get two readings the same, until John points out that snow melts: the legs of the tripod have been steadily sinking as the sun rises. But even when we dig down to soil, it isn't good; soil, I learn, melts too. Finally, John takes over on setting the location of the house and does it first time, no problem. I'm humiliated, even after he tells me he worked for two years as a surveyor in the Arctic. That night, while the guys go off to have a beer, I take a hot shower and have a long talk with myself in the pink cubicle that is now my only private space.

Captain Beefheart hasn't arrived by day two, but John has managed to borrow shiplap for footings. Since this house is meant to be portable, we'll build simple concrete pads for support posts instead of a continuous wall and, because there are no Redi-Mix trucks here, we'll mix and pour the concrete by hand. While Ray and John collect water from the town pump, Dave and I start; he digs while I cut lengths of shiplap.

"How big are you making the footings?" He's not so much digging as using the crowbar to gouge chips out of the frozen earth.

"Sixteen inches." I slice off a length and pick it up to use it as a pattern.

He frowns. "Don't you think that's too small?"

"No." I bite my tongue before I can say, "Why do you ask?" Maybe there's something you haven't thought of and Dave has, the critic inside my head whispers. I grit my teeth and saw off another length with unnecessary force. Then I set the saw down hard on the sawhorses and raise my head.

We're in the middle of a small valley set amid low rolling hills under the dome of a magnificent blue sky. There's a lightness to the air here that makes me feel as if I might float. A low whoop grows inside me, and my arms naturally rise to embrace everything. This place makes me feel large, damn it, and I'm not going to let myself be diminished.

—

Still no sign of Captain Beefheart, and Ray reminds us nervously that we're two hundred miles from the nearest supply store. It costs ten cents a pound to ship in materials. When I wonder out loud where we'll get gravel for the concrete, Ray says we'll use whatever's lying around the site.

"But the Building Code says it has to be clean!" I know he knows this.

"This is what we have, so this is what we use." So we do.

Just as we've begun to figure that Captain Beefheart has indeed taken off with our supplies, he arrives—four days after he promised to be here.

"Had to spend a few days with my son in Prince George," he explains, "to finish haying."

We spend the rest of that day sorting and piling and after supper, to celebrate, we all go for a beer with Jan at the Dease Lake pub.

It's a surprisingly cozy room lined with dark wood, its walls lightened with pictures. There's a dartboard and a pool table and round tables covered in green terry cloth. The clientele of leather and grizzle and beard feels friendly, though the Native people and the whites, I notice, sit mostly at separate tables. Still, as we sip our beer, Jan calls out to everyone, inviting them to join us. She introduces me with, "She's a carpenter, she's the boss!" until I ask her to stop.

Actually, half my job here feels like timekeeper. I have a fantasy that if I didn't tell our little crew to stop at the end of each day, they'd work until they fell, nap on the spot, chew a hunk of raw moose and get back to work. Ray never objects when I call for lunch or coffee—he just never calls it himself. Dave, I suspect, doesn't dare.

Ray makes my life easier by helping to lift and generally facilitate when I need another hand—or head. He's plumber, electrician, heating man and general consultant, but he never wears a tool belt. He tells me that a wall fell on him on one job, and he's been a supervisor ever since. I can see now how his bad hip slows him.

Dave works hard and never complains, no matter how tedious or unpleasant the job. I appreciate his willingness, though he rapidly falls into a dour aspect broken only by an occasional dry observation. Much later I'll find out he's left a wife and two kids behind. John, when he isn't flying, is general nail pounder, but he's on the site less and less as the days grow longer and his plane is more in demand. When he's around to take his turn at cooking, it's usually moose meat from his freezer—tough but good, like spicy roast beef.

I've become a quiet person here. There's little need to talk on site except to give direction, and when Jan talks, I'm so fascinated by the stories she tells, by her matter-of-fact approach to hardship, that I forget to answer. There isn't much talk of women's liberation up here, and the sex roles are very, very clear. I begin to see why. One day, when John is asked to fly supplies into a gold mining camp where the ice he'll land on is "probably okay," he flies. The roads here are ice in winter and blinding dust in summer. Northern conditions, Jan says, make most city men "perfectly useless." Suddenly "macho" makes sense. These men have more physical courage than I do, so if they want to strut about after doing the dangerous, difficult work that has to be done to survive, they've earned it.

Funnily enough, here in the land of macho, I feel more accepted as a construction worker than I ever did in the city. Men don't blink when I'm introduced as a carpenter, although the women show a lot of interest in how I got here. Jan says it's because most people who come north don't fit down south any better than I do. Also, there's only one other carpenter in Dease Lake, so people don't have the luxury of caring who you are. If you're useful, you're welcome.

There's a total of two very small food stores in town, and everything in them is expensive. Milk is $1.60 a litre. There's very little that's fresh, either fruit or vegetable. One night when it's my turn to cook I have a craving for cauliflower. Ray says, "Good luck." But I find it, the only cauliflower in town, brown and withered to the size of my fist. It costs the outrageous sum of five dollars, and I buy it without a second's thought.

⌐

After footings are poured and the support posts up and braced, we're ready for floor joists. When Dave begins to pack over two-by-eights, I find Ray where he's tinkering with the water tank.

"That floor needs two-by-tens."

"Just use the two-by-eights." He hardly looks up.

"I'm sure the Code calls for tens."

"So we're going by the book, are we?" It's John, standing behind me.

"Didn't you bring your Code Book, Ray?" I offer to go check mine.

"Use what we've got."

What's the matter with these guys? Don't they want to do this right? "What about the inspector?" I'm bringing out the big guns now.

John starts to chuckle. "Ah, inspection! Who wants to be the inspector?" he booms.

I maintain proper forewoman decorum. "It says in the Book," I start again, and John lets loose with a belly laugh. When Ray looks up, he's grinning.

"So you're saying scrap the Book?" I ask, just to get things clear. Ray and John both roar. I smile, then giggle, then the three of us are bent over, clutching our sides with laughter while Dave, watching from a safe distance, continues to carry two-by-eights.

For the rest of the job, it's a favourite joke: "Kate, what does the Book say now?" followed by a loud guffaw.

I begin to see another way of thinking. Here in Dease Lake, where the nearest building inspector is five hundred miles away and the lumber yard a mere two hundred, people get the necessary work done with available materials, under impossible conditions, against all rules. One day when we take the afternoon off to drive to Telegraph Creek, John points out a house resting on sheer rock, with no foundation at all. It's over one hundred years old and still lived in.

I had thought my job would be on the hammer and saw, but it turns out Dave and John do most of that. Mostly I plan and direct, so when there's a mistake, like the joists get laid out wrong, everyone knows it's my fault.

"Not a mistake if you catch it!" John roars kindly.

And in fact, the house is going up smoothly. Ray is counting; we have less than three weeks left.

—

Since we arrived, light has been rushing in upon us. The light here is yeasty with air beaten into it, has weight and density so it splashes over us, leaving me giddy. I walk around buoyant as a rubber ball. Even better, Jan reminds me that this far north, by late June there'll be hardly any darkness at all. Seeing how much there is to do, and not wanting to waste the wonderful light, I've stopped worrying about hours of work. We work from nine until five, then from eight-thirty until six, then seven, then eight. Only the cook leaves one hour early to have dinner ready when the rest of us get home.

On Saturday night there's a dance in the community hall, and Jan comes by to give Dave and me a ride. The hall is a small log structure built like a child's idea of the perfect house. The front door sits dead centre in the front wall, with a rectangular window on either side. The night is cold, yet half a dozen people, mostly men, stand outside in shirt sleeves, smoking and talking. Several are Native, with hair that swings like sheets of dark water when they move.

"Tahltan people," Jan says when I ask.

The only light inside comes from the brightly lit bar, framed in low-hanging glasses and a shiny black counter that reflects jars full of olives, lemons and shockingly red swizzle sticks.

People begin to sway—men in cowboy boots and embroidered shirts, women in their best blouses—even before the musicians strike up the first tune, an energetic mix of country, western and rock and roll. Heedless of the fact there's no caller, men hoot and do-si-do and swing their partners until the women are shrieking. If we were in Vancouver, I'd ask Jan to dance, but I don't see any other women together, so I hold back. Then I spot Dave, tapping his foot.

When I approach him, he solemnly nods and moves onto the dance floor, where he extends his arms and begins a series of long slow spins and swoops. I dance around him, relieved to move my body in something other than the prescribed motions of construction. When Dave takes a break, I dance on my own, then with a smiling man who appears in front of me for a while and just as smoothly disappears. Dave comes back and we dance again until the lights suddenly come on and the band stops. I blink. There are only a few people left in the hall, and Jan isn't one of them. Dave and I walk back to the motel beneath a fantastically starry sky, in silence. Five hours later we're eating breakfast, ready for another day. I don't feel tired at all. It's only as we start laying plywood that I remember.

"Today is Sunday," I say, standing up suddenly. "Ray, this is our day of rest!"

But Ray has disappeared into the shed, and neither Dave nor John looks up. I settle for declaring to empty space that this will be a short day, and at five o'clock, with the sky still bright, I pronounce us finished.

It's the first week of May, and the weather is shifting fast. Each day starts with a blizzard. I go to work in long underwear, arctic boots and down jacket, and by 3:00 p.m. I'm in shirt sleeves and gumboots, sweating under a blue sky with a light wind—until a cloud passes in front of the sun and I dive for my jacket. Sweet and sour weather; it's exhilarating. One day when a sudden snow storm wipes out a beautiful sunny morning, John complains.

"Don't feel bad," I call from the top of a wall. "It keeps the bugs down."

"That's the spirit," he says heartily. "You're fitting right in."

At work I'm a laser focus of concentration, but the rest of the time it's as if my body is a child's playdough, being worked into new and unrecognizable shapes. It's the place that makes me feel this: the light, the size, the space. That, the tightness of our time schedule and my yearning for privacy. Every night from the moment we stop work, I look forward to the shower so I can hide inside a stream of steaming-hot water. Then Jan tells me about Telegraph Hill.

Telegraph is the highest spot in town. Although the official sign at the base says "No Trespassers Allowed," everyone goes there; it symbolizes the anarchist independence I like about this place. Whenever I can borrow the truck I drive halfway up the hill, then make the short, steep climb to the peak. Its name comes from the fact that in the 1860s this was the route planned for an overland telegraph trail that would link Europe and North America. The route was abandoned when a cable was successfully laid across the Atlantic instead. The top of the hill is a broad table scattered with white quartz, so even when it gets dark I can see where to step, as if walking on stars. I sit on a large, flat rock watching night come, wind flailing the hills blue-black. I begin to think about my writing; I'm nearly forty years old and have always said I wanted to write. If I'm going to get serious, maybe I'd better start soon?

The only thing still to worry about on the house is the roof. This is the first roof I've been solely responsible for, and on the morning we're to start, Ray is sick. When we get to the site, Dave and John look at me sideways when I ask them to count sheets of roofing and two-by-sixes for rafters.

"Why?" John asks.

You can't do this! the critic begins inside my head. "So I'll know we have all the materials we need."

"It's a waste of time. You're not going to get more now anyway. Why don't we just start building?"

"Because I don't want any surprises," I reply, my jaw tight.

I figure out lengths and angles and lay out the first two rafters. I set Dave to cutting rafters and show John how to mark the top of the walls while I recheck my figures. There's a moment of panic when I realize that, because of where the porch roof falls, we're short one rafter. I call Ray.

"If you can't hide it, highlight it!" Ray orders. I change a few measures and we're ready. The first two rafters fit perfectly, and we set up a system: Dave and I stay up top to nail while John passes up rafters, nails and anything else we need. By six-thirty the rafters are finished.

"There's your house," I say to John as the three of us stand back to admire its skeleton against the backdrop of hills. The hardest part is done.

The next morning we apply bright orange galvanized metal roofing, and by noon, less than three weeks after we dug the initial excavation, we're finished—as carpenters say—to closure. I give a whoop and go for my camera to take pictures of a horizon I'll not see from this exact angle again. Ray, who's back, declares a roofing party.

"We'll fly to Alaska," he announces. "Supper in Wrangle."

By three o'clock, John is turning the little Cessna on the runway. By now I've taken several flights with him, but I'm alarmed each time at how light the plane is, how small its propeller—two wooden paddles that must keep several of us, plus the plane itself, safely in the air. But people here take plane rides the way people in the south take buses.

As we lift above the lake, John explains that the name of the river we're going to follow, the Stikine, comes from a Tlingit word meaning "great river." It begins as a slit in the earth, the Grand Canyon of the Stikine, that runs furious and fast, then opens up as the country around it grows increasingly rugged. For two hours we fly over a kaleidoscope of forest, water, mountain and glacier. When the Stikine finally spreads out into a sandy delta, the great blank space on the map that was once northwestern British Columbia begins to fill in for me.

From the little airport in Wrangle, we hitch a ride into town in the back of somebody's pickup ("Limousine service!" John announces), then separate: Ray wants something at the hardware store, John needs supplies for the plane, and Dave wants to see the petroglyphs—ancient stone carvings—on the beach nearby. I have something else in mind.

I smell it the instant I step inside: a lush smell that makes me think paradise isn't hot at all, but cool and green with the perfume of fresh fruit and vegetables. After luxuriating in the supermarket's produce section—sniffing, squeezing, admiring the brilliant colours—I buy strawberries, bananas, avocados, three jars of my favourite hot sauce and three bags of corn chips. Then I have to hurry; John can only fly in daylight, and Ray has promised seafood. We eat quickly and hitch another ride back to the airport. John takes a different route home, following a carpet of gold laid down by the sun setting behind us. I can't decide which is more beautiful: mountains swathed in the peach and rose of sunset, or the glaciers stretched over them, fascinating for the dark shadows of crevasses, like hairs on a belly, marching down to the darkness of a sudden drop into space.

—

The next afternoon, as Dave finishes up the insulation and I finish back-framing, the knuckles on my left hand begin to itch. I put on gloves, thinking it's a delayed reaction to the fibreglass I helped lay earlier, but by suppertime, red itchy welts cover the backs of both my hands. By bedtime, my nose is itching and the welts are also on my neck.

"It's the cold," Dave says.

I make a face at him. I worked in a T-shirt all afternoon.

"You've been sniffing too hard at those strawberries you bought in Wrangle," Dave tries again, "and they're coming through on your face."

"Or you're working yourself into a frenzy," John says. He's right. My emotional skin is stretched tight as a drum. Earlier that day when Dave made some small comment at work, I came up with an excuse to go to the truck so he wouldn't see me cry.

Ray keeps his back resolutely turned as he watches a tiny black-and-white TV perched on the kitchen table. I raise my voice.

"Ray, I want a day off."

"Can't take it, eh?" His eyes never leave the television. Pure Ray, but I'm too tired for this macho shit.

"I've been working ten hours a day for three weeks, with one whole day off. It's time for another one."

"Anyone else want time off?" he asks, eyes still on the TV.

I have to give it to him: the man has an uncanny skill at getting people to work harder. I know Dave and John are as tired as I am, but they rise to his bait and deny wanting time off. Cowards.

The next morning while the others make breakfast and leave, I barely open my eyes. There's a brief kafuffle as Ray—with no phone at the new house—comes back to check on John, whose plane has gone through the ice at a camp in the mountains called Muddy Lake. Once he confirms that John and the plane are okay, I drift off to sleep again. At noon I emerge from a thicket of dreams, drugged with fatigue and the luxury of being alone. I eat a leisurely poached egg with the last of the Wrangle strawberries, then retreat to the couch with a blanket and the book I've barely touched since I arrived. I sleep again.

When I wake, I head outside and follow the high-pitched whine of a vacuum cleaner to find Sandy, who runs the motel with Al, on her hands and knees under a bed. She'd mentioned leading a women's keep-fit class.

"Drop-ins are always welcome," she says now. "Thursdays at seven, at the school." Before I leave, she says, "If you ever want a break from those guys, come over here. I have some very nice bubble bath."

Even the offer is a gift.

These days at work we take our coffee breaks in luxury, sitting on lawn chairs in the middle of a half-drywalled living room. During our morning break on Thursday, I announce I'm leaving early that day to go to keep-fit class.

"What for?" John asks.

"Yeah," Ray says. "Don't you get enough exercise here?"

"Why don't you just admit you're going to miss me?"

"No way," Dave says. "We'll finally get some peace around here!"

Just before seven, the three of them wave goodbye as I climb into the truck, promising to pick them up at eight-thirty. As I pull off my steel-toed boots at the gym door, I'm relieved that the women already lying on the floor are dressed like me in blue jeans and cotton shirts. There's no big-city Spandex here. I feel a shot of joy to be in the company of women again. Sandy turns on her boom box and motions for me to find a space on the floor.

Instead of the aerobics I expected, Sandy leads us in stretching and isometric exercises, and slowly, I sink into the structured calm of—Bach! I don't have to be tough here, or make a thousand decisions. I don't have to pretend my body is the same as any man's, nor do I have to ignore it. I let some soft part of me out of hiding and surrender, at ease, letting Sandy's voice tell me what to do.

One afternoon when John is away flying and Dave's left early to fix supper, Ray and I take a quick coffee break. I mention that Dave feels bad about taking so long to put on the exterior siding, and the next day I over-hear Ray casually mention to Dave how good the siding looks. Ray pushes, but he's easy to work for, never gets angry at mistakes or dumb questions. He's relied on my judgment for this job as much as I've relied on his.

We're into the pure-pleasure parts of building now, cruising toward the end. While Dave and John finish drywall, I hang doors and use Ray's por-table table saw to cut windowsills and returns, baseboards and moldings.

"If John wants cupboard doors and bathroom tile, he can do it him-self," Ray announces at coffee two days later. And with that, the house is finished.

Ray declares a housewarming/farewell party for the next night, before he and Jan and I drive south together. A couple of the women in town have asked if I'd stay on to do renovations for them, and I've decided to say yes, as soon as I can get back here with my own truck and my dog, who's currently being minded by my roommate. John has invited me to stay in the new house with him and his family, who will arrive soon, in exchange for finish work on the house. Ray thinks I'm crazy, but I'm not ready to leave the north just yet.

We spend our last morning furiously touching up paint and plaster, and our last afternoon cutting up salami and laying out crackers, chill-ing the beer. The whole town is invited. Sandy brings over her boom box and dance tapes, and Jan arrives with a fresh-caught salmon. I find tin foil and lemon, and before long there's a salmon cooking in the brand new oven. The place is alive with people dancing in the living room and talking in the kitchen and drinking everywhere.

It's still bright outside when I notice Ray out front, methodically packing the truck. I put down my beer and go outside to help, with Dave right behind me.

"No," Ray insists. "You young people go have a good time. I'm al-most finished." Dave hesitates, then leaves. For a minute Ray and I stand looking at the house squatting beside the runway in its cedar cladding and bright orange roof. It looks as if it's been here for years.

"Want to know something, Ray?"

He's comfortably silent beside me.

"This is the first house I ever built from scratch."

"I know," he says. "You did fine."

17

finding gold

When I come back to Dease a few weeks later, I do a roof repair, lay some linoleum, build a deck and finish up the house for John and his wife. Pretty soon, people are asking about my plans for fall, especially after I rent a tiny log cabin.

Several times, I've heard John and some of the other pilots mention a woman, Theresa Bond, the only female bush pilot in this part of the country. Mostly they talk about her mistakes, but I get suspicious after the third or fourth time I hear the same story, about how her plane once hit the corner of a floathouse.

"Doesn't she do anything right?" I ask John.

"Sure, but we wouldn't want to admit that, would we?"

One day when I'm laying the lino at John's, a woman wearing a blue coverall strides into the house. She's the tiniest adult I've ever seen, and I know immediately.

"Theresa," I say, standing up and putting out my hand.

She takes one look at my tool belt.

"Kate."

And I wonder for just a second what they've been saying about me.

Over the next few months, Theresa and I see each other maybe three times, but it's immensely comforting just to know she's out there. One of those times, I ask how she gets the full barrels of diesel onto the plane.

"Leverage," she says. "And balance." Of course. I've learned the same.

Another time she asks, "Have you heard what they say about me?"
I'm not sure how honest I should be, but her face is calm, open.
"Yes," I say. "There's some version of that about all of us."
She nods.

Toward the end of July, I start a job for a woman named Alyson, who's asked if she can work with me. We're laying two-by-four decking and I've just shown her how to space the boards the thickness of a spike. We're heads down, hard at work, when a voice as clear as Alyson's beside me says, "I don't need to do this." I stop work and listen, as if tuned to an internal radio station.

"This is my work now," the voice says. "I do it for myself. I don't have to prove anything to anybody." It's simple, unquestionably clear, and it throws everything into a new light.

"What?" Alyson asks, seeing me sit up.

I look at her. "I was just thinking, I'm a carpenter."

"You sure are," she says, and we go back to work.

Perhaps it's nothing to do with the fact that I've raised my rates to match those of the only other carpenter in town, but after Alyson's job, I get no more work. Just as I'm thinking I'll have to leave, John points out that piles of lumber are being loaded into cargo planes at the airport. It's all going to a gold mine so deep in the bush that flying is the only way in. There's always been gold there, John says, but with rising prices, it's suddenly worth taking out.

"And where there's wood"—his eyebrows go up—"there must be carpenters!"

That night we go over to the motel office, where John's arranged a radio phone call with Jeff, the camp superintendent. He pushes buttons on an imposing black box, then we wait for an operator.

"Everyone in the north listens to this thing," he warns. "It's our favourite entertainment. If people forget that, the rest of us get to hear some humdinger conversations. A miner who's been in camp for six weeks can fair burn the wires up."

He shows me how to press a certain button when I want to talk and let go of it to listen. "Say 'Over' when you finish talking."

"Hello." It's Jeff, the camp superintendent.

"Hello, I'm Kate." But I've forgotten to press the button, so he hasn't heard me.

"...understand you're a carpenter looking for work. What experience do you have? Over."

"I've done a lot of house renovations and some high-rise..."

"Come in please. Over."

"Press the button when you talk," John reminds me.

Finally I get it straight. "I've done a lot of renovations and some high-rise work," I say. "Over."

The camp is called Muddy Lake, and the job involves replacing the original miners' tents with winterized cabins. Jeff pays room and board plus ten dollars an hour if I bring my own tools.

"We work ten hours a day, seven days a week," he says. He can use me for one week, starting immediately.

The next morning, John has already been flying for a couple of hours when he comes back at seven to pick up my tools, Ruby and me. The plane's already loaded. He frowns at the spotty clouds, then we take off with a small roar. Ruby cringes on my lap. I tell her, "This is going to buy you more dog food!" Then I ignore her to look out the window. Below us are dark rolling hills polka-dotted with trees, but the landscape turns vague as more cloud moves in. A small plane like this flies by sight, not instrument, but I'm only mildly worried. I trust John, who now hands me a road map.

"See if you can find those two lakes there below us."

Just as I'm warning him I can't find my way around on the ground, forget the air, he dives through a small window in the clouds, and I blink at a suddenly vibrant blue skyline.

Within a few miles, the landscape has changed dramatically. There are no trees. Here and there on the hills I can see the dark mouths of caves, then a small lake. At the last second, a crack opens in the mountain directly in front of us, and we're flying along a narrow valley running parallel to a silver scratch of river.

John concentrates on dials and gauges, checking the wind as he gets ready to land on what looks like a gravel path. Three people stand beside the runway.

"Either my new passengers are special," John says, "or you're a major curiosity. That's Jeff himself, the camp manager, down there. He never comes to meet me."

The Cessna has barely rolled to a stop before John's out. He's all hurry and efficiency as he unloads a small mountain of supplies and the two passengers clamber in. Cargo delivered, job done. And I suddenly understand he's going to leave me here.

"Have fun," he calls as the door bangs shut and the plane starts rolling.

Jeff and I haven't said a word. By the time I stop gazing after the Cessna and look around, Jeff is almost out of sight over a small rise with my tool belt over his arm. I grab my tool box, shoulder my pack and hurry awkwardly after him with Ruby close behind. He's headed toward the lake a few hundred yards away, where the two of us and my unhappy dog pile into a small black dinghy. My brain flashes disaster headlines: Boat Engine Refuses to Start! Pair Mauled by Maddened Grizzly! But the dinghy starts instantly and we're off. I steal a look at my new boss. His lips are pursed and he frowns as if lost in thought. He looks over his shoulder to where the dinghy is headed, then at his watch, then over his shoulder again. He reminds me of the Mad Hatter in *Alice in Wonderland*. That makes me Alice.

The lake lies in a bowl formed by a circle of mountains. A fine spray of waterfall spews rainbows over the water to our left. As I pull my jacket closer, Jeff yells over the noise of the engine, "We're at 4,600 feet here." He looks again at his watch and twists the handle of the motor, and the little boat spurts forward. He doesn't speak again until we're tied up at a small wooden dock and my gear is piled on the deck.

"Cook shack's straight ahead," he says. "You might as well get a coffee. The others will take a break any minute." He looks at Ruby—"Dog stays outside"—then turns and disappears.

The cook shack is a big tent. I make two trips up to the door, leaving my gear in a small hall at the entrance. Inside, lengths of brown burlap drape from every second ceiling panel, and the canvas at the sides comes down to meet four-foot-high sheets of faded grey plywood and a plywood floor. A row of picnic tables covered with white-and-brown checked cloths runs up one side of the tent, a low counter along the other. The space feels like a cozy cave and, even better, smells of fresh-baked bread.

At the far end of the cook shack are two stoves, two sinks, shelves loaded with pots and pans and cooking supplies and, in front, a long table where two aproned women peel potatoes. When the door bangs shut behind me, they look up.

"You're the lady carpenter!" the blond one exclaims, setting down her knife and wiping a hand on her apron before extending it to shake my cold one. "I'm Annie," she says, "and this is Bobbie. Are you hungry?"

"I wouldn't mind some toast," I say. It's been only a few hours since breakfast, but it feels like another lifetime.

"Help yourself," Bobbie says. "Coffee and tea are always hot."

On the counter to my right are seven jars of cookies labelled in cheerful script: peanut butter, spice, checkerboard fudge, raisin, chocolate chip, gingerbread and sugar. Another jar of licorice and one of trail mix stand beside a cardboard box full of chocolate and granola bars and under the counter is a shelf lined with juices, soft drinks, puddings, canned fruit, loaves of bread, crackers, peanut butter, honey, jam and cheese. A four-slice toaster stands close by, a large breadboard on either side.

I feel better already. I fix myself a toasted peanut butter and honey sandwich and coffee, rich with cream, before sitting down at the table nearest the cooks.

"Is it true you were the first woman in the union?" Annie's blond hair is pulled back in a pony tail, but wisps float loose around her face.

"Just in Vancouver." I wonder how they know I'm in a union at all. John must have been talking.

Annie raises a fist. "Yeah, women's lib!" We're all laughing as the door opens and heavy boots announce the building crew.

The first few men in the door have dark faces, eyelids drawn low over black eyes that glance at me quickly, then look down. With a start, I realize the crew I'm going to work with are Native men. Several sit down at the table behind me, talking in low tones. The last man stomps up the corridor between tables calling dramatically, "Where's our new carpenter?"

"Kate, meet Simon, the town mayor."

Simon sits down directly in front of me and leans forward over his crossed arms. He wears a blue plaid jacket, and his thick black hair falls forward over lively dark eyes. "I guess you'll fit right in if you can find the food this fast. I'll show you around."

My dog is clearly relieved when I reappear at the front door, following Simon. Muddy Lake camp lies behind the cook shack and consists of twenty or so faded tents built—like the cook shack—with canvas tops and plywood sides and floors. They're lined up on each side of a grassy space filled at one end with building supplies. Across the centre sags a droopy net.

"Volleyball," Simon explains as we walk. "For when the wind stops blowing, which is almost never. You'll notice the cold, too. We'll have snow here by September, and in October it can be minus forty." He points to where the two tents farthest from us are being replaced with a cabin of two-by-four and plywood. "That's what you'll be building."

He points again, to a tiny shed nearer the lake. "And that's our sauna," he says. "The big tent behind the cook shack is the TV room."

I follow as he continues to walk down the row of tents. "Once you guys get the cabins up, I put in the electricity and stoves. I'm the electrician, plumber and heating man. And mayor," he says modestly, glancing at me for a second and grinning. I smile, but I can't seem to find my tongue.

Simon stops at a tent on our right and pulls open the plywood door. Inside are two wooden bunks with a rough plywood table at the foot of one. The canvas top creates the same warm light as in the cook shack. Travel posters are tacked on the lower half-walls.

"Welcome to the luxury suite." He lays my pack on the table and puts the tool box beneath it. "Most of us don't have a table, just shelves. There's no one else in here for now, but you could have a roommate any time. Our little town is growing by leaps and bounds," he says proudly. "There are fifteen of us already, and miners coming in as fast as you can build their houses."

When he closes the door, telling me to ask if I need anything, I see a poster on the back, of a woman naked under a wet T-shirt. The slogan on the bottom reads, "Curaçao es Corazon!" Curaçao Is All Heart. It doesn't seem as offensive as it might have been in the south.

I buckle on my tool belt and step outside, Ruby at my heels. There's no one in sight. For a moment I have the strange sensation of being able to see, as well as hear, the wind. Across centre court, Jeff strides toward me followed by a heavy-set, white-skinned man with a cloud of red hair and a red beard, wearing a carpenter's apron.

"Larry'll tell you what to do," Jeff calls over his shoulder as he disappears again.

It's me who breaks the silence. "Would you like me to tear down one of these tents?"

"Not this one."

"I'll be her helper."

Larry ignores the man who's just spoken, now standing behind him. "You can start on that one." Larry points to the tent beside mine. "Use the same floors."

That's all the instruction I get. I watch him walk away along the row of tents, looking as if he should be on the high seas with a pipe clamped between his teeth.

"What do we do first?" I ask Wayne, who's now my helper.

Wayne says nothing, just walks over to the tent and begins pulling the old tarp off the frame. I watch him, then do as he does, loosening the nails that tie the tarp down, carrying it to the grassy centre where we fold it into a tidy bundle. Jeff has told me that all materials here—even the plywood—will be recycled. It's something I've never heard of.

"What do you call this patch of grass?" I finally ask. Silence may be okay for a while, but talk is one of my anchors.

"We call it the Common."

"You mean, like in old England?"

"Yeah. The Common."

"It goes along with Simon being Lord Mayor, right?"

Wayne nods. They have a sense of humour in this place.

The old framework of two-by-twos and plywood is surprisingly sturdy. At first I try to carefully pull rusty nails from the dry lumber, but they've been here too long and the heads keep breaking off. I'm trying not to swear, but "Shit!" slips out.

Wayne grins at me and goes back to work. He's using the sledge-hammer technique, prying a gap between pieces of lumber with the claw of his hammer, then just bashing them apart. Once, after an older carpenter watched me neatly tap at a stubborn piece of wood for too long, he ordered, "When in doubt, use violence." I follow Wayne's lead and soon the little structure—all except the floor—is flattened.

The platforms are eight by twelve. Wayne and I lay the first wall out on the deck but when I start to square it, he gestures for me to stop. His voice is flat, the sounds pushed out with the slightest pause in between.

"We put it up. How it is. Plywood goes on. However it fits."

A little voice inside me whispers, that's not what it says in the Book! "But it's not square."

"Neither's anything else."

"Is the floor level?"

"Nope."

"So why don't we level the floor?" I'm not feeling entirely sure of myself here.

"Never did before."

We spend the next half-hour bracing and shimming to get the outside edges of the floor level. Wayne goes along without comment, but when I stand back to admire our work, I see that the centre now dips alarmingly, and there's no way to crawl underneath to raise it. When I insist we carry on anyway, "by the Book," Wayne helps me square and sheath and lift, then stands patiently as I inspect the huge gap along the floor where the walls—precisely square—fight to follow the roll and wave of the floor. At the corner where the walls meet is a hole big enough for Ruby to crawl under. Wayne stares at the offending walls, then turns to me.

"Don't fit."

I burst out laughing. "Okay, let's do it your way."

We build the last two walls entirely à la Wayne, putting up the stud frame regardless of plumb or level and slapping on plywood wherever it fits. The walls fly up in record time.

After a spectacular lunch of homemade soup and mountains of grilled cheese sandwiches, pickles and cake, it's time for rafters. I'm a slow learner; I have an overpowering impulse to build these rafters by the Book. So how do we cut rafters on a frame that's neither square nor level?

Farther down the row, the other crew are tearing down their second tent. On the other side of us, Larry, working alone, is starting walls.

"So how would you build the rafters?" I bear down on "you" a shade more than is polite.

"Measure 'em."

"Eh?"

He repeats, "Hold 'em up. Measure 'em. Can't do it on the ground." His calmness is infuriating, not to mention the fact that he's been right every time—so far. I set up the sawhorses, but when I turn to give him the end of my measuring tape, Wayne's gone. At first I'm annoyed; then, feeling like a kid whose parents, by incredible good luck, aren't watching, I cut the first set of rafters "on the ground," using the mathematical formulas I've been taught. They don't fit.

When Wayne reappears, he climbs to the top of the walls and extends his hand for the tape measure as if he'd been gone thirty seconds and not thirty minutes. Without a word I climb to the peak of the little house and pass him the tape. By five o'clock, when Larry calls for clean-up, we're finished.

As Wayne puts the tools away, Larry walks over. "Sorry you had to get Wayne. He wanders a lot."

"It was okay," I say, feeling an urge to defend my silent helper.

Larry looks skeptical. "None of them are that great. That's why I work alone."

"Them," I assume, is the Tahltan men. "I prefer to work with someone,"

I tell him, and am relieved when Jeff's voice interrupts from across the Common.

"Larry, show Kate the dries."

Larry waves vaguely in the direction of the cook shack. "There."

I'm asking his back, "What's a dry?" as Simon walks by and offers to show me.

The dry—to the right of the cook shack—has two doors to it. Inside one are two large sinks with a tiny mirror above one of them, and two hand-built showers with barroom-like plywood doors. Simon assures me there's tons of hot water at Muddy. Hearing that, my worries about short doors fade. Behind the second dry door is an empty room with a stove in the middle and clotheslines like spider webs across the ceiling. Three-inch spikes ring the walls every foot or so.

"Once they start mining," Simon says, "the miners will leave their clothes here to dry out. They have priority, but for now, anyone can use it."

In my tent, I unpack and head toward the shower with clean clothes over my arm, underwear discreetly stuffed in my pocket. Today I built a tiny house right up to the rafters, and I like working with Wayne. He gives me room, as if we're forming some sort of pact: I won't judge you, if you won't judge me.

—

I wake in pitch dark in the middle of the night, needing badly to pee. The outhouse is a long trip in the dark from my tent to the far side of the Common, and I've forgotten to bring a flashlight from Dease. Actually, I'd packed one, then taken it out of my bag, thinking surely the camp would have lights. I've also forgotten slippers. The dirt under my bare feet is fine as plaster dust, and cold. I walk carefully toward the back of my tent, aware we've piled plywood full of rusty nails back here. When I get to where I figure is far enough, I pull up my nightgown and squat. As I stand again, I hear a faint crackling above my head and look up.

The sky is bright with stars, but there's an odd light—not dawn—and the stars seem to be fading. I shake my head. The sky is definitely rose-coloured now, then green, then yellow, colours sliding down as if some gypsy woman in a bright silk skirt dances over our heads. I watch until the colours lighten, until I'm so cold I can't stand still and am forced to dance with the northern lights.

The next night at supper I meet Linda, the fourth woman in camp.

She's quieter than the cooks, and works in the office as camp administrator and First Aid attendant. I also meet Dale, the catskinner. A catskinner, I find out, drives the heavy equipment made by Caterpillar. Dale's main job is to keep the road to the mine mouth open in spite of all attempts by the mountain to shut it down.

"With seventeen switchbacks on that short climb to the mine mouth," he explains, "the company should be planting grass. Hold the soil better. Good for the road, good for the mountain sheep." He straightens his broad shoulders and places a large, square palm on the table in front of him. "But I'll do anything the company wants and do it the way they want it. Hell, I'd pickle my mother if they told me to. But they should plant grass up there."

On my third night I wake abruptly. There's no sound: no voice, no bird, no motor, not even wind. As if I'm in a vacuum. The bedside clock says 1:30 a.m., and I feel stuck in a half-dream, almost desperate. It takes a while to get back to sleep, and in the morning it crosses my mind I could get claustrophobic here. People in camp talk about "getting out," ask, "How long are you in for?"

One of the joys of this place, though, is the time I have to read—all evening, every evening. As I lie in my bunk with *Hanta Yo*, a historical novel about the Teton Sioux before contact with whites, I read, "Throw out everything; regard yourself as a newborn."

Every morning I get up at six and head to the dry to wash my face. I've stopped looking in the tiny mirror above the sink. It doesn't seem important. Then I waltz to the cook shack and order breakfast—anything I fancy, from porridge to bacon and eggs, to cold cereal or pancakes. It's better than a restaurant, because it's made in front of you by two cooks who beam as they pass it over. I work from 7:00 a.m. to noon, with a ten-minute break at ten. At lunch there's hot or cold food, in the afternoon another break. We quit at five, followed by the luxury of a hot shower for as long as I like. Hot water has been my refuge in the north, my spiritual solace, my solitude.

Every meal here is abundance, but suppers are extraordinary. On a typical night we might have veal cutlets, fresh (fresh!) tomato slices, carrot salad, onion salad, baked potato with real chives and bacon, sour cream and butter and frozen peas. For dessert there might be home-made chocolate cream puffs, chocolate ice cream and fresh fruit, plus the usual cookies. Quantities are unlimited, and there's no shame in going back for more. Which I do. Frequently. The cooks take this as a compliment, and you'd hate to disappoint them.

I've begun to understand what a deal men get when they have a traditional wife—the kind who stays home to clean and mind kids and, best of all, cook. In the city, when all I wanted to do after work was to come home and collapse, the chores felt endless: laundry, cleaning, ironing, planning meals, shopping, cooking. Here I do none of that. I never realized how much brain space simple planning takes. Living in a camp is like having two wives and two mothers, with everything but my laundry done for me. I don't even have to worry about my dog, who I now rarely see except at bedtime when she curls up at the foot of my bed. With those sad beagle eyes, she's quickly become the camp mascot; everyone pats and feeds her, especially the cooks.

In the city—even in Dease Lake—there was always a gap between work and home, but here, that line doesn't exist. We're a fast-growing town of twenty-six people, with a solitary volleyball net for entertainment. The camp is officially "dry," though I know John occasionally smuggles in a bottle for some miner. At the last minute, I'd packed my birth control, but in a town of twenty-two men and four women, I quickly decide I don't want to underscore that I'm a woman. The thought of sex feels like a match in a keg of dynamite, though I'm not sure who's the match, who's the dynamite. Usually I'm an affectionate person, but in Muddy Lake all of us keep careful boundaries. I don't even hug the women, as if any reminder of the physical beyond food, work and sleep might let loose a landslide. Pete, the diamond driller, has told me some people come on purpose to a camp like this to dry out, but for the most part, our personal lives are a subject on which we all keep a firm silence. We ask no questions. We live in a cocoon of the present.

One night, over supper, we discuss how Jeff and Eric, the geologist, always look serious. Eric insists it's because he has to work so hard for Jeff; Jeff says it's because he has to figure how to make enough money to feed Eric. Who could have known, he asks with a mournful countenance, that such a skinny guy could drive the company almost to bankruptcy? Eric toasts that possibility with a third helping of potatoes.

"I'll win the Muddy Lake Food Olympics, no problem," he assures us and glances at Jeff who—as usual—is checking his watch. "And Jeff will be the official Muddy Lake Timekeeper!" Everyone laughs. At dinnertime, we let go. Something in this place, it seems, has to run wild, and better it's our imaginations than anything else.

Our only other outlet is volleyball. One of the miners, an ex-fisherman, hand-knotted the net that stretches across the Common. On the first evening that the wind dies, every available soul is hauled from their tent and dropped on one side of the net or the other for a noisy game, then we all go back to the cook shack for hot milk and Ovaltine. As I look around the table, I feel as if I've lived with these people for years. It's been four days.

I know the bosses are watching me. Heck, the Tahltan guys are watching me too—as I watch them. In their eyes, I see how slow I am— the girl—how careful. Too careful. The northern way of doing things is to make do, so when Jeff shows me a clothes rod made—in the absence of proper dowelling—from a triangular slice cut from a two-by-four, I'm impressed.

At the end of the first week, he asks if I can stay.

"We've decided to replace the cook shack and drys as well as the miner's tents, and I need carpenters."

I'm happy to.

"Planeload of miners coming in today," he tells me the next morning, and asks me to build a bed in the newest cabin. Simon, waiting behind me to install the stove, decides to help by throwing together a concoction of two-by-fours with a half-inch plywood mattress while I finish hanging the door. When I complain his bed is ugly, he laughs.

"Who's going to sleep here, the Queen?"

And I have to admit, after he lays down two inches of foam, it's comfortable enough. After that, and the triangular closet rod, I start "building northern," as Simon calls it. The miners who are beginning to fill the camp thank us profusely for their tiny new houses with the spanking white insulation and two plastic-covered windows. No one ever complains that a wall is off-plumb.

—

After seven days I still know nothing about my helper, Wayne, except that his home is a town near here and he's never built anything before. He seems happy to follow my lead, and when I get the sense he's about to disappear, I reschedule the work coming up, planning jobs that are easy for one person until he comes back.

I move out of my tent into one of the new cabins and, on the first night, leave my door open as I lie on the bed reading. It's cold outside, but the diesel heater hums, and the clothes I've just washed in the camp's ancient wringer washer hang over every parallel surface, drying. From outside comes low laughter from a miner's cabin and the creak of a door in the omnipresent wind. The only other sound is the unending whine of the diamond drill, nosing for gold high on the mountain. Earlier today we heard the warning whistle and looked up to see clouds of red smoke and earth belly out from the mountain's peak, then, several seconds later, the sound of the blast. They're driving the shaft deeper.

My new house smells of fresh lumber and Twilight Grey paint. Simon hauled over my table—"Save yourself work when you can!"—and on it, at the foot of my bed, sit my alarm clock and a small photo album propped open. I'd written to a friend in Vancouver that the only thing I miss here is flowers, and he'd sent this small treasure by return mail—a bouquet of photographs. Even their names are poetry: aster, dahlia, rose. The colours explode like firecrackers in this world of brown and gold, where I huddle in my down jacket and huff warm breath onto my frozen fingers.

Glad no one from home can ask why, I've taped "Curaçao es Corazon!" on the back of my new door. In the city I'd be insulted, but here she represents for me the barest female presence, in more ways than one.

Bobbie's dark head appears around the door. I'd almost forgotten that, with the camp filling up, it's time to double up in the cabins: Bobbie and I are now roommates. While she spreads out her sleeping bag and settles in, we talk. Her hours are 4:30 a.m.—5:30 if she works an extra hour of prep at night—until 10:30 a.m. Then an hour off before lunch shift from 11:30 until 1:30. Dinner runs from 3:00 until 7:00 or 8:00—whenever the kitchen is set up to start over again the next morning.

After she leaves to finish up her shift, I do some figuring. That's twelve hours a day of work, and she gets only six dollars an hour. If a carpenter—who makes ten—has a bad day, the rest of us cover and nobody notices, but if a meal here was five minutes late, or the food not delicious and plentiful, there'd be a riot. And just two women to feed us all. So which work is "more important"?

⌐

One morning, as a miner brushes his teeth at one sink, I floss mine at the other. The large hulk of Pete, the driller, stands in line behind me waiting to use the sink next. Suddenly I flush with embarrassment. I've never flossed my teeth in front of a boyfriend, let alone a stranger! But Pete behind me is calmly picking at his own teeth, unselfconscious.

"I feel like I have twenty-two brothers."

"Guess that gives me four sisters."

August 13, and at the mine mouth five hundred feet above, there's a heavy blizzard. Here below, it's a balmy 5° Celsius with 55-km winds roaring through.

Lately Wayne's been wandering off more often, so I'm not surprised when one morning I go to work and Scott, one of the other Tahltan guys, stands ready to be my partner in Wayne's place.

Building with Scott feels more like a partnership. He makes me laugh. And he sings. When I tell him I've never met anyone who knows the words to so many songs, he says, "I know," modestly, then grins an abandoned, immodest grin that makes me laugh again.

Pete, who works from 7:00 p.m. to 7:00 a.m. seven days a week, has invited me up to watch what he calls the rig—the diamond drilling machine—in action. So one night after supper I go with Eric on one of his regular checks up the mountain on the Kubota, a three-wheeled motorcycle. As we climb at a steep angle, I strain to get a clear view over his shoulder. Halfway up, he yells over the sound of the machine, "Watch for mountain goats." We see them once, far away but thrilling, grazing on a high patch of grass.

Within minutes, Muddy Lake lies in a tiny bowl below us, mountains rising sharply in clean cones all around. As we near the drilling rig, the distant noise I've been hearing for days grows louder, harsher, until, by the time we reach the drill site, I have to stuff my ears with tissue. The site is beautiful, but the rig, perched on the edge of a cliff like some plywood animal sticking its back end out into the wind, lets loose great smelly farts of steam and gas.

Inside the rig I watch Pete raise the drilling machine to remove a long cylinder of rock ("core," he says) about four inches in diameter. He lays this cross-section of the earth into a narrow wooden box he labels with where and at what depth he found the sample, then piles it with several identical boxes in the corner. Mohammed, the assayer, will use these earth tailings to map exactly where gold is hidden inside the mountain.

—

Everyone keeps a calendar. Today is Day Eleven.

—

The brilliant gold of mornings is ever sharper, but by afternoon it's overcast and cold between bursts of watery sunshine. I routinely wear a down jacket, wool hat and wool gloves with the fingers cut out. Occasionally I remember the warmer south, but every time I think of leaving here, especially after supper when I want nothing more than a quiet conversation and someone to hold me, it's balanced by the pleasure of waking up to brilliantly clear air, the magnificence of mountains, and Scott's good company. He's started to tell me stories about his life on the reserve, and now routinely sits beside me at meals. Lately he's also taken to chewing tobacco. When I ask if he likes it, he replies, "No. Tastes like ashes." But he keeps chewing. "Spitting's the best part," he says. "If I chew, I can spit, and tell certain people to go to hell."

Every day now, he asks me to tell him what I read the night before in *Hanta Yo*. One afternoon, he says, "I ever tell you about the first Indian who wins a gold medal in the Olympics?" I shake my head. "When he takes that medal, he holds up an eagle feather. To show the Indian peoples he is Indian and the peoples can be proud."

The story gives me goosebumps. That night I dig out my pink dress-up blouse from the bottom of my pack and wear it to supper. I'm no beauty queen: my face is tanned, my nose is red and chafed from constant blowing, my eyes are bloodshot from the wind. No matter how much cream I slather on, my skin is as dry and scratchy as the Styrofoam we use for insulation, yet I'm feeling a powerful combination of soft and strong, as if my confidence and pleasure at work allow me to assert the other parts of myself, too. Working here, I'm accepted as a woman and a carpenter both, more balanced than I've ever felt in my life.

The light the last few mornings has taken on a deep, tawny shade, and when I step outside my cabin, snow is low on the hills. I stay in touch with home by radio phone, regularly calling Kevin back in Vancouver and sometimes—when his dad answers—having a chat with John.

—

George, who holds the Muddy Lake Olympic medal for Silence and leads the third crew, is the only Tahltan man who's been hired as a carpenter, not a helper. We've never worked together, never even talked, but a wordless respect has grown up between us. The houses George works on go up quickly, and lately we keep running into each other, like the time he hears me asking for a new saw blade and brings it himself.

One day as George is working on the house next to Scott and me, Larry orders him to start on a different one. George doesn't move a muscle. His head stays exactly level with Larry's, but his eyes lower a fraction and he's gone, quick as that, his spirit disappearing in a non-gesture that says as clearly as if he'd shouted, "I don't have to do what you say."

I turn back to work, hit something with my hammer to keep from laughing out loud. It's the cleverest write-off I've ever seen. When I glance over at Scott, he's watching me. Then he spits off the roof, on the side where Larry is standing. The next morning, George climbs up to where Scott and I are working, pulls out his hammer and proceeds to work with us as if this has somehow been agreed on. And perhaps it has.

A few days later, Scott gets sent up the mountain to apprentice as Pete's helper. I miss his company but it's a promotion for him, and George becomes my regular partner.

I happily chatter through our days together. George responds with a nod or a word and sometimes not at all, but I always have the sense of being listened to. One afternoon as we're working on the roof of the new storage shed, I spot the new guy as he comes out of the cook shack. He'd flown in the night before, a bigwig from down south who'd spent breakfast telling everyone, in a voice like stone on glass, how we ought to be running this camp.

"I don't like that guy," I say softly, watching him walk up the Common. "Talks too much."

George turns to look at the newcomer, then at me.

"Maybe he just wants to be heard," he says.

His insight takes the breath out of me.

One day George and I discover that I'm making one hundred dollars a day and George only ninety. I don't say maybe it's because I have a carpenter's ticket, and I try not to think maybe it's because I'm white. I tell him the difference is probably because I brought my own tools, but in his quiet way, George lets me know he doesn't think it's fair.

"I do the same work as you."

Until this moment we've been easy partners, passing the lead back and forth, but a few minutes later, when I wonder out loud how we should handle something, George says, "You're the carpenter. You're paid more." And waits. He has reason to be mad, but I want to get this roof finished by supper. So now, for the first time in my life as a woman construction worker with a male partner, I seize the lead. A few times I start to panic; can't subtract, hear the voice of doubt—then I pull myself together. You know how to do this!

George knows I like him, that I'm having fun, and after a while he gets into it. At five o'clock he looks around and says, "Worked good, eh!" High praise. And that is the day I finally, fully, believe I am a carpenter.

⌁

The last structure to be replaced is the cookhouse, something all of us have looked forward to building. Because it's the biggest building on site, I assume Larry will use all three crews, but when we report for work, he sends everyone to do a major cleanup and, as I move off with them, calls me back.

"You and I will start the cookhouse."

I tell myself it makes sense for the two of us to lay it out first—but I can't help feeling I've betrayed the others.

The foundation, roughly thirty feet by ninety, is pieced together with twenty-foot logs, twisted and bowed. As Larry holds up one of them for me to slip shims under, it drops on my hand. I yell, but he doesn't react until I scream his name. Then finally he lifts it so I can pull my hand loose. My fingers aren't broken, just badly bruised and cut, and it's an accident, I'm sure, but that night Scott comes to my cabin and announces darkly that the others are certain Larry did it on purpose.

"He's mad you like us."

Larry announces a second, then a third day of cleanup for the rest of the crew. He and I, he announces, are going to build a shack for the miners up at the mine portal. I'm intrigued at being able to see the mine face but the rest of the crew are left down below, and again I feel like a traitor.

Larry and I finish the foundation, then eat our lunches, perched on the edge of the mountain overlooking the valley.

"When you phoned Jeff about work," he says between mouthfuls, "he was desperate for carpenters."

"The story of my life."

Then he starts talking about the Tahltan men.

"I'd fire them all in a minute if there wasn't so much politics about it. You ask them to do something and they act like you're criticizing them. Can't take orders. "

It sounds sort of like what men have said about me, but nothing I could say will make any difference. "I'll be right back," I tell him to stop this useless conversation, and head down the road and around the corner before I pull down my jeans and squat to pee.

But Larry isn't finished. As we start framing walls, he says, "I don't understand what some women are on about, wanting change. My wife doesn't understand it either. She doesn't want anything different from what she has."

I'm busy banging spikes. I tell him absent-mindedly that some men don't want women in construction, on principle. "And not every woman wants to be a carpenter. I just think we should have the choice."

"So what do the men do, the ones who don't want you there?"

"Some big stuff, like not hiring you, but mostly small stuff like not talking when you walk in the lunchroom, not telling you what the plan is, never allowing mistakes. That's what gets to you."

"But of course they'll take a man on the job before they'll take a woman. Women are more trouble. They need special facilities, Johnny-on-the-Spots."

I look at him, astonished. Haven't I just gone around the corner to pee? I've finished the end wall and start to lay out the long front one.

"What are you doing?" he asks sharply. "Build the short one next."

He's the foreman. I build the short wall. Then I notice him raising the other one.

"Larry." I hesitate because it's so obvious. "If we raise both short walls, it's going to be hard to get the long one raised inside them."

He interrupts, ordering over his shoulder, "Grab the other end of this."

There's a small dropping feeling in my chest. "Did you hear me? This is going to make it hard…"

"I heard you." He continues to raise the short wall, and after a moment, I grab the other end. I rationalize it to myself, but the rest of the day goes like that, and by afternoon, I've begun to make mistakes; I cut a header wrong, hit my thumb, trip over the lumber. A few days ago, working with George, I'd felt better than ever about my skills. After one day with this guy, I'm back in the same place I was a year ago. How long is it going to take before they can't do this to me anymore?

After supper, Scott knocks at the cabin door. I'm sitting cross-legged, telling Bobbie and Annie on the other bunk about my miserable day. Scott sits down and listens.

"I'm glad that happened. Now you know why we don't like that guy."

"But it's different," I say. "He was trying to make me feel bad because I'm a woman."

"That's the same."

And I hadn't recognized it. Prejudice is prejudice.

⟶

That night I dream I'm on a hilltop leaning against a funeral bier. But no one has died. It's Sleeping Beauty's bier, and on it lies a child, almost comatose. Behind us is a tree, gnarled and twisted, without bark or leaves. All the beautiful parts—leaves and flowers and bark, the places for birds to play and wind to sing—have been stripped away. There's no beauty to this tree, just a dead core, what carpenters call heartwood, and even that is barely hanging on.

On my way into breakfast the next morning I pull Jeff aside and tell him I want to go home. His eyes flash regret before he looks away.

"You sure? You're welcome to stay here over the winter. You fit in well in the camp."

I'm sure. I tell him I've just had a letter from a friend who's about to get married in Vancouver, which is true. But there's another reason. The dream has shown me there's a price to be paid for what I'm doing. I've developed some parts of myself—the powerful trunk, strong branches—but I miss other parts, the creativity. It's time for integration. I have no idea how I'll do this.

I'll leave tomorrow on John's next scheduled flight. I tell Scott and George, then Bobbie and Annie, who invite me for a sauna that night. After supper, the men stay indoors while Bobbie, Annie and I fetch buckets of cold water from the lake, then leave our clothes in a pile outside and share the steamy heat of the little sauna. I remind them that I spent my first few minutes in camp with them. I feel quieter now than the woman I was then.

After the sauna, when I go to Scott's cabin to say goodbye, he asks, "You gonna leave me that book?" Ever since he's gone up the mountain, he hasn't heard any more of *Hanta Yo*.

"I'll leave it for you," I say. "And thank you." I close the door behind me, hoping he knows how deep that thank-you goes.

I've been here twenty-one days.

18

what's in the flesh

Back in Vancouver, I quickly pick up several renovation jobs. A week after I get home, I'm in the truck when I hear on the radio that a woman pilot in northern BC has just crashed into a lake, killing all five passengers. The pilot is the sole survivor.

It's Theresa Bond.

For the next few days I listen to every broadcast and buy every newspaper, trying to get up the courage to write. In my mind I compose a thousand letters. Finally, I phone.

"It was my worst fear," Theresa tells me, "that I would survive and my passengers die." She was thrown out the pilot's windscreen. "I can't swim, you know. When it happened, I thought, I'll just hold my breath and it will all be over. But I couldn't."

Her voice is strained but calm. "People have been so kind, the women especially. They keep me sane. One woman who's been widowed twice told me, 'I forgive you, and when you forgive yourself, you'll be whole.'"

We're both crying. John Hill, she says, was the only pilot who visited her after the accident.

Theresa will never fly her own plane again, but over the next few years I'll hear through John how she's put her life back together, become unit chief of the local ambulance paramedics. In May 1990, though, there'll be another report. Theresa, in a small plane with another passenger, a child, was being flown to Smithers for a paramedic's meeting when the pilot hit a mountain and all three died instantly. She was forty-two years old.

Jacqueline asks me to work with her again. It's one of the biggest jobs either of us has ever done, a reno for two women on the west side of town. We'll rip off the front of the house and build extensions front and back, plus a series of complex, multi-levelled entry decks. The architect is a big-personality lesbian, happy to work with women carpenters, though her first loyalty is to her clients. I like it that her priorities are as clear as ours.

I also like it that I can focus on building while Jacqueline does the contracting. Our crew is Gina, Jacqueline's apprentice, and, when there's extra labouring work to do, Kevin, who's now sixteen. So why, with this sweet set-up, do I head back into the all-male fray when the union calls me with a short job?

Something about the men draws me back. I miss their stupid humour and their rough caring. I miss the camaraderie that means we keep an eye out for each other's safety but otherwise ask nothing of each other, especially emotionally. I even miss the rough and tumble of their constant challenge. With women, it's easier. Too easy, perhaps. And there's so much emotion. I can hide with the men, keep a clear distinction between life on the job and life off it. Mostly, with the men, I can be some part of me that I can't be anywhere else.

I say this to myself, and in the next minute wonder if it's true. I thought I felt most whole when I worked with women? Didn't my dream show I needed more female company? But I want the whole dance. I want both. I want balance.

The union job is to build a downtown high-rise. Most of it is fly forms, but some parts still have to be hand-formed. My partner, Lorne, and I are assigned to form up the elevator shaft, building what will amount to a multi-storey concrete tube around which the ironworkers will later erect the rest of the structure. Our work space is jammed with formwork, lumber and screw jacks, and one of the first things we do is to pack twelve-foot-long, six-by-six-inch posts from where the crane's left them to the elevator shaft, to reinforce new concrete as it's poured above. We're protected most of the time by guardrails, but to place the posts on the shaft side, someone has to hang out over space. Lorne is six-foot-four, 220 pounds, and can scoop up one of these hunks of tree "as if it were a penny," as the rigger says. But when Lorne can't squeeze into the narrow space where we have to fit them, the foreman sends someone out to buy a safety belt small enough for me, and I'm the one who hangs over the elevator shaft, wedging beams into place.

I can't figure Lorne out. He does the minimum he has to to get along with me. That's hardly grounds for complaining, yet a small warning bell rings.

Lorne can't understand why I want to be a carpenter. One day when he launches into it again, Rod, our labourer, is nearby. As Lorne repeats his question, Rod mutters, "Yeah, bloody stupid."

I'm used to not feeling welcome, but from Rod I feel something I've never felt before: pure hatred. He says almost nothing to me directly, goes to great lengths to avoid me, laughs as I pass. When I ask for something, Rod pretends not to hear. When I ask him again, he says, "Speak up." Once, he outright sneers, his face curling into a mask that scares me. The rigger tells me Rod has been with this company so long he's the only non-foreman who routinely eats in the foremen's shack. Increasingly, after breaks, it's Rod who tells Lorne what we'll be doing next, keeping his back turned so I can't hear. Increasingly, Lorne doesn't pass the information on to me. I'm left to guess, all the time trying not to notice the smirk on Rod's face.

Labourers have a lot of power. They're called "semi-skilled," but they set the pace; if your labourer doesn't bring the materials you need when you need them, it not only slows the job but makes the carpenter look bad.

By now, we've started dropping plywood platforms into the empty space between the concrete core and exterior walls at every third floor. "Fourteen stories is too far to fall," the foreman says, though we all know even three stories can do serious damage. The plywood platforms, being temporary, are only wedged into the space, and there's a three-inch gap between plywood and concrete.

One day I'm working alone on the concrete platform. I've been bent over in the same position for some time, vaguely aware that Rod is working three stories above me—cleaning, probably. For a moment, I swing my upper body to do something at my right side, and in that instant, a heavy steel pry bar plummets straight down and lands, point first, in the exact place my spine had been seconds before.

When I look up, I see Rod's eyes watching through the three-inch gap.

"Rod! You could have killed me!" I'm shaking.

Rod says nothing. I hear him going down the scaffold stairs and I am certain this was no accident. Rod will do anything to get me off this job. I take a few deep breaths until I stop shaking. You asshole,

Rod, I think to myself. There's no way in hell you're getting rid of me now.

"You're on, Rod!" I call out loud.

That is Wednesday. On Friday, they lay me off. No reason; they've just hired a new carpenter, and there's clearly lots of work. I figure Rod finally got through to the foremen in the lunch shack, convincing them I was no good. And maybe it's for the best. I might have died there, because I was never going to give Rod the satisfaction of quitting.

To fill in time until the next job, I take an upgrading course at the school where I did my apprenticeship training, now renamed the British Columbia Institute of Technology (BCIT). The teacher often talks to me about construction—what jobs I've done, where, who I've worked for—and one day he asks if I can take over a pre-apprentice class for a couple of weeks, just until the regular teacher gets back from holiday.

It's never crossed my mind to teach construction, and I have no idea how you do it. I'd be the first woman instructor in this department. How will male students react? But it's only for two weeks, and the money's good, so I say yes.

I make it through the first morning, covering theory, and after lunch take the class—sixteen men—into the yard. When I was a pre-apprentice here with Jude, a heavy-duty mechanic instructor named Tom took the two of us for coffee one afternoon.

"There are no women instructors at this institution," he'd told us. "I want to see you back." Now I hear a deep male voice bellowing, "Sister!" It's Tom, his arms flung wide in greeting.

Two weeks at BCIT becomes a full-time job. Over the two years that I teach there, the students—sixteen of them every six weeks, all male except for a single woman—are, without exception, respectful. They think it's normal that a woman teaches them how to build. John and I are living together again and have decided to get married. So why am I so unhappy? At home, I revert to stickies on the fridge that assert, "I can do this job!" Still, I cry every day as I drive to and from work. I can hardly explain it, even to myself. When a friend asks what's the matter, I tell her I don't know what's in my students' heads. It sounds silly, even to me, but it's true. Teaching gives me nothing to lay my hands on—hands that have never looked so good, so smooth. That don't look like my hands anymore.

One day at lunch in the staff room, another instructor asks if I can carry a two-by-ten.

"If I only carried one," I shoot back, "I'd be fired." It's the right answer, and the others laugh. I've one-upped him, and now I deliver the coup de grâce. "How about you?" He's older than me, longer off the tools, out of shape. It's a little twist of the knife.

I've done it lots of times, this kind of talk. But suddenly I've had enough. I'm tired of being the odd one, the exception, even if it's just a joke. I want to be normal. A few days later, I give my notice, then agree to stay for two more weeks when the department asks me to cover for another instructor still on vacation.

I decide that since I've already quit, I'll teach this final class the way I wish someone had taught me. To hell with blending in; I'm going to be a woman carpentry instructor. So I'm more chatty than ever and ask lots of questions. When they don't know the answers, I challenge them to give a best guess, to use their own minds instead of waiting for me to feed them answers.

"This is how it'll be on the job," I tell them. "Another problem, another decision. How are you going to solve it?"

When we come to the most boring section, on materials, I set them up in teams and send them on a treasure hunt. When we get to building stairs, I take them outside to walk up and down different sets of stairs, getting a five-, a six-, then a seven-inch rise into their bodies as well as into their heads. I'm showing them the tricks of the trade, how a carpenter thinks. I let myself have fun.

When I tell my class I'm leaving, two separate groups of students ask me to stay. I'm moved by the fact that young men want to learn the way I did, but it's time to go. I'm tired of having to prove myself. And I have another reason for leaving: I want to write. In fact, writing is now the one thing on earth I most want to be doing.

On my last day in Dease Lake, the only day I'd worked indoors all summer, I'd tuned in to a Labour Day program on CBC Radio in which a man was telling animated stories about his work in a factory. I'd never heard anyone talk with such passion about blue-collar work before.

When the man finished, the host said, "That was Tom Wayman, reading his poems."

Poems? But he'd sounded just like any guy in a lunch shack. If those were poems, then what I'd been writing in my journals all these years might be poetry too. Back in Vancouver, I'd picked up a leaflet for something called the Kootenay School of Writing and ended up in Tom Wayman's class. I could hardly believe my luck. Even better, when the course was over Tom had invited me to join his writing group, the Vancouver Industrial Writers' Union. That group became my hand-hold. We met once a month to read our latest poems aloud and to make them better. We also organized readings at a local café, and it was at one of those that I first read my work in public. Hearing my name called, standing up, moving forward, mounting the stairs to that tiny stage and reading words raw from my unguarded heart was the single most ter-rifying thing I had ever done. But it quickly became the most exciting, the most urgent.

After leaving BCIT I earn my living as an administrator in Con-tinuing Studies doing liaison between Simon Fraser University and the labour movement, but my heart is in poetry. I send my poems out to literary magazines, where some of them get published. My first book, *Covering Rough Ground*, about my life in construction, wins an award for the best book of poems by a Canadian woman in 1991.

I'm a good administrator but something in me is dying; I have to write, and in order to make as much time as possible to do it, I go back to school for a master's degree in creative writing, then start teaching creative writing myself. Finally I'm exactly where I want to be, doing exactly what I want to do. So I'm surprised by the strength of my reac-tion when someone introduces me one day as a woman who used to be a carpenter. It stops me dead. Used to be?

For fifteen years that work engaged every fibre of my being. Not just all my senses, though those too, but all my muscle, all my skill, all my intelligence, and very quickly all my love. I can no more leave that behind than I can leave my body behind. I realize it only as I say it.

Not used to, I tell them. I'll always be a carpenter.

acknowledgements & thanks

As I was writing the final draft of this story, I visited my parents, now living in Nanaimo, British Columbia, for Mother's Day, and that night returned to Vancouver. The next day my dad had a bad fall and was taken to hospital. He was eighty-six years old and his health had been failing for years. "Nothing serious," the doctors said. "He'll be home in three days."

Three days later, my sister phoned to say Dad could not be woken. He was taking no food or water and recognized no one. My father was dying. I took the next ferry to Nanaimo and went directly to the hospital. He lay deeply asleep, not moving. I took his hand and sat down by his bed.

He stirred under the blue hospital blanket, then shook his head and very slowly opened his eyes. He said, "Is the book finished yet?"

I was amazed. In the last few decades my father and I had reached a deep peace together, but he knew my story included him, and some of our many battles. An hour later, as I gently let go of his hand, he pulled himself awake with the same intense effort, opened his eyes and said, "Never forget how proud I am, how much your mother and I love you and your brothers and sisters."

At his funeral, I planned simply to read a poem by Mary Oliver, "The Summer Day," which ends, "Tell me, what is it you plan to do / with your one wild and precious life?" But as I stood up, I realized I wanted to tell the people there—people who had loved my father for all his complexity and difficulty—something more.

"From the time I was twelve years old," I started, "my father and I fought." Some of my siblings flinched. Was I going to start ranting at his funeral? And truly, I didn't know what I was going to say. I was working it out as I spoke.

In all that time, I continued, Dad never put me down, never demeaned me. He took me seriously, argued back, kept talking. What I was learning in those times, I told everyone, painful as our battles were, was that I had a right to be heard.

I didn't know it until finally I said it out loud: "My father taught me courage." And my mother was beside him every inch of the way. Thank you, both.

The events in this book are reported as accurately as I could make them, with the help of my extensive, often daily, journals. In some cases I have changed names to protect people's privacy. My thanks to the Canada Council for the Arts and the British Columbia Arts Council for financial support in the writing of this book. Also thanks to Mabel Dodge Luhan House, especially Karen Young and Liz Cunningham, for a residency in Taos, New Mexico, that allowed me to do concentrated work on the final draft. Short excerpts from this work appeared in an earlier form in *Slice me some truth: An anthology of Canadian creative nonfiction* (Wolsak & Wynn) and in *Walk Myself Home: An Anthology to End Violence Against Women* (Caitlin Press).

There is a group exercise that involves allowing yourself to fall backward, trusting that you will be caught by the people around you. My fifteen years in the trades felt like that. At the critical moments there were always people there—sometimes people I didn't know, or knew only for that moment—who caught me with a look, a gesture, a kind word. There are far too many to thank by name; many of their names I don't even know. But I thank them, and all the men who worked with me, who wondered about me and worked with me anyway, who taught me what they know. Deepest thanks to the men who took a chance and hired a woman, especially Ray Hill, Ted Bowerman, Jac Carpay, Heron Douglas and Colin Boyd.

I also worked with women; not many, but they were the more precious for that. Warmest thanks to the women in and around Vancouver Women in Trades, without whom I would never have lasted: Sharon Boudier, Judy Doll, Suzanne Girrard, Carlyal Gittens, Rose Hilton, Janet Lane, Marilyn Lantz, Joan MacArthur-Blair, Alice Macpherson, Janet Pinder, Lynn (Lark) Ryan, Carolyn Sawyer, Ann St. Eloi, Alison Stewart, Heather (Watt) Tomsic, and others, including Sue Doro, Susan Eisenberg and Molly Martin in the US where parallel events were (and are) taking place. Special thanks to Marcia Braundy, Chryse Gibson, Jacqueline Frewin and Gina Horrocks. You gave me heart as well as backbone.

Thanks to Susan Booth and Carol Brookes who taught me why "Why?"; and to Herta Buller, Mary Anne Paré, Beverley Richardson, Glenys Sherer, Elke Stoll and Wendy, who kept me healthy.

Three successive writing groups were vital in their encouragement and feedback for this book that's been twenty-five years in the making: the Vancouver Industrial Writers' Union (Brad Barber, David Conn, Glen Downie, Kirsten Emmott, Phil Hall, Christine Hayvice, Zoë Landale, Sandy Shreve, Pam Tranfield, Mark Warrior and Tom Wayman); SexDeathandMadness (Kath Curran, Cynthia Flood, Christine Hayvice, Bonnie Klein, Joy Kogawa, Claire Kujundzic, Sheila Norgate, Carmen Rodriguez, Tana Runyon, Sandy Shreve and Thuong Vuong-Riddick);and particular thanks to the Memoirsistas (Heidi Greco, Joy Kogawa, Susan McCaslin, Elsie Neufeld and Marlene Schiwy) for their care.

Thank you to Heather Bishop for permission to quote from "A Woman's Anger" (www.heatherbishop.com). For work on the manuscript, thanks to Keith Maillard and the students of my UBC writing class who commented when it was struggling to stay fiction. Also to Helen Cook, Heidi Greco, Vlad Konieczny, Sandy Shreve, Betsy Warland, and especially Barbara Pulling for inspiration, challenge and genius at helping me cut and shape. Thanks to Ted Bowerman, Ryan Braid, Marcia Braundy, Jude Campbell, Tom Clarke, Don Crane, John Hill, Art Kube, Barbara Kuhne, Janet Lane, Brendan Lillis, Louise Mandell, Wendy Munroe, Valerie Overend, George Park, Annabelle Paxton, Liora Salter, Bill Sealy and Bill Zander, for technical advice and corrections. Any errors, technical or otherwise, are my own and I'd appreciate hearing about them.

I'm very grateful to the writers of the Creative Nonfiction Collective, whose discussions of "what is creative non-fiction?" and the place and techniques of memoir were helpful at many points. Memoir writing has been a surprisingly emotional exercise, and for help over the rougher patches, for their words and writings, I thank Brian Brett, Joy Coghill-Thorne, Karen Connelly, Keith Harrison, Susan Juby, Zoë Landale, Cathy Ostlere, Molly Peacock, Bonnie Schmidt and John Terpstra. Many thanks to the women at Caitlin Press, especially Vici for your commitment and hard work. I thank Fear and I thank the Wild Women who guided and guarded me through it all.

My deepest love and thanks to my family for unfailingly supporting my choices no matter how strange. And what can I say that would express my thanks to my husband, John Steeves, who went through this with me—both the life and the writing about the life, the rational and the irrational—and to my son, my chosen child, Kevin Steeves, who eventually was proud that his mom drives a truck. Thank you for rewarding me with a daughter-in-law, Alina, and a granddaughter, Zlata Sophia. I have been so lucky.

⚊

In the forty years since changes to Human Rights legislation first allowed women to enter trades, the number of women in trades in both Canada and the US has stayed at roughly 3 percent. Clearly, we still have far to go. For a partial bibliography and resources on women in trades in Canada and the US, see www.katebraid.com.

photo credits

PHOTO BARRY PETERSON

Kate Braid worked as a receptionist, secretary, teacher's aide, lumber piler, construction labourer, apprentice and journey carpenter before becoming a teacher of construction and creative writing. She has taught creative writing in workshops and at Simon Fraser University, the University of British Columbia and for ten years at Vancouver Island University (previously Malaspina University-College). She is the writer, co-writer and editor of several books of poetry and non-fiction. Kate lives in Vancouver.